T0144487

Cellular Network Planning

RIVER PUBLISHERS SERIES IN COMMUNICATIONS

Series Editors

ABBAS JAMALIPOUR
The University of Sydney
Australia

MARINA RUGGIERI
University of Rome Tor Vergata
Italy

HOMAYOUN NIKOOKAR
Delft University of Technology
The Netherlands

The "River Publishers Series in Communications" is a series of comprehensive academic and professional books which focus on communication and network systems. The series focuses on topics ranging from the theory and use of systems involving all terminals, computers, and information processors; wired and wireless networks; and network layouts, protocols, architectures, and implementations. Furthermore, developments toward new market demands in systems, products, and technologies such as personal communications services, multimedia systems, enterprise networks, and optical communications systems are also covered.

Books published in the series include research monographs, edited volumes, handbooks and textbooks. The books provide professionals, researchers, educators, and advanced students in the field with an invaluable insight into the latest research and developments.

Topics covered in the series include, but are by no means restricted to the following:

- Wireless Communications
- Networks
- Security
- Antennas & Propagation
- Microwaves
- Software Defined Radio

For a list of other books in this series, visit www.riverpublishers.com

Cellular Network Planning

Marcelo Sampaio de Alencar

Federal University of Campina Grande, Brazil

Djalma de Melo Carvalho Filho

NUTES, State University of Paraiba and UNIFACISA, Brazil

River Publishers

Published, sold and distributed by:
River Publishers
Alsbjergvej 10
9260 Gistrup
Denmark

River Publishers
Lange Geer 44
2611 PW Delft
The Netherlands

Tel.: +45369953197
www.riverpublishers.com

ISBN: 978-87-93519-22-0 (Hardback)
       978-87-93519-21-3 (Ebook)

©2017 River Publishers

This book is dedicated to our families.

Contents

Preface

This book is the result of many years of research and consulting in the area of cellular planning.

Over the recent years, few books have been published covering all the subjects needed to understand the very fundamental concepts of cell planning. Most books which deal with the subject are destined to very specific audiences. The vast majority of the books introduce the subject at a very basic, or technical, level, or are destined to an academic audience.

This book begins with an introduction to the subject, covering conventional and contemporary wireless systems. Spectral allocation and the frequency plan are discussed, along with the essential characteristics of wireless systems. The design of mobile cellular systems includes cell planning, traffic and channel problems.

The book presents a review of existing models, considering both green field dimensioning and network expansion strategies, and discusses multi-objective optimization and base station deployment based on artificial immune systems. It also discusses a cost-effective base station deployment approach based on artificial immune systems, and introduces the modified MO-AIS algorithm.

Some of the technical topics discussed in the book include an introduction on mobile cellular network basics, a discussion of the evolution of mobile cellular systems, a definition of the mobile communications channel, a presentation of several mathematical propagation models, the cell planning process, along with the green field dimensioning and the possibility of network expansion, and a discussion on cost-effective planning strategies. The book also introduces the concept of cell planning using voronoi diagrams.

Marcelo Sampaio de Alencar
Djalma de Melo Carvalho Filho

Acknowledgements

I am grateful to my colleagues and students at the Institute for Advanced Studies in Communications (Iecom), Federal University of Campina Grande.

I am also indebted to my family for the patience and support during the course of the preparation of this book.

I also acknowledge the support of Mark de Jongh, of River Publishers, and his staff, especially Junko Nakagima, who provided the review, suggestions and guidelines to prepare the book.

Marcelo Sampaio de Alencar

A large number of people have contributed to my evolution both as a professor and as an electrical engineer. First and foremost, I should thank my students, both undergraduate and postgraduate for inspiring me every day, especially the ones of Computer Networks, Introduction to Computing and Computer Organization and Architecture.

I would also like to thank professor Edson Guedes da Costa, professor Wellington Santos Mota, professor Washington Luiz Arajo Neves and professor Marcelo Sampaio de Alencar for allowing me to pursue and develop a career in research in Electrical Engineering when I was still a university student.

I am also very grateful to Professor Gisele Bianca Nery Gadelha, Professor Hamurabi Medeiros and everyone else at UNIFACISA for giving me the opportunity to start off as a university professor and be part of a successful team. Thanks also to professor Misael Elias de Morais for inviting me into the Center for Strategic Technologies in Health (NUTES), believing in me and boosting my career as a professor, a researcher, an electrical engineer and a lawyer. Thanks to everyone at UEPB and the Computing Department for the encouragement throughout the writing process.

This book would not have been possible without the support of the Institute for Advanced Studies in Communications (Iecom). I would also like to thank everyone involved in the development of the manuscript, the editor, the reviewers and the publishers.

Djalma de Melo Carvalho Filho

List of Figures

List of Tables

List of Abbreviations

3G	Third Generation Cellular Systems
3GPP	Third Generation Partnership Project
AMPS	Advanced Mobile Phone Service
ARQ	Automatic Repeat reQuest
AWGN	Additive White Gaussian Noise
BER	Bit Error Rate
BH	Busiest Hour
BRMOA	Binary-coded Multi-objective Optimization Algorithm
BS	Base Station
BTS	Base Transceiver Station
CCS-7	Common Channel Signaling Number 7
CDMA	Code Division Multiple Access
CMOIA	Constrained Multi-Objective Immune Algorithm
COST	Coopération européenne dans le domaine de la recherche Scientifique et Technique
D-AMPS	Digital Advanced Mobile Phone Service
EOM	Economic Optimization Model
ESN	Electronic Serial Number
ETE	Equivalent Erlangs
ETSI	European Telecommunications Standards Institute
FCC	Federal Communications Commission
FDD	Frequency-Division Duplex
FDMA	Frequency Division Multiple Access
FM	Frequency Modulation
FSK	Frequency Shift Keying
GIS	Geographical Information System
GoS	Grade of Service
GMSK	Gaussian Minimum Shift Keying
GPRS	General Packet Radio Service
GPS	Global Positioning System
GRC	Group of Radio Channels

GSM	Global System for Mobile Communications
HF	High Frequency
HSDPA	High-Speed Downlink Packet Access
HSPA	High-Speed Packet Access
HSUPA	High-Speed Uplink Packet Access
ISDN	Integrated Services Digital Network
ISI	Inter Symbol Interference
ITU	International Telecommunications Union
LTE	Long-Term Evolution
MAHO	Mobile-Assisted HandOff
MAI	Multiple Access Interference
MIMO	Multiple-Input and Multiple-Output
MIN	Mobile Identification Number
MISA	Multi-objective Immune System Algorithm
MO-AIS	Multi-objective Optimization Algorithm based on Artificial Immune Systems
MOCSA	Multi-Objective Clonal Selection Algorithm
MSC	Mobile Switching Center
MSC-H	Home center office
MSC-V	Visited center office
MT	Mobile Terminal
MWVD	Multiplicatively Weighted Voronoi Diagram
NMT	Nordic Mobile Telephony
NTT	Nippon Telegraph and Telephone
PDC	Personal Digital Cellular
PMD	Principle of Majority Decision
PSD	Power Spectral Density
PSN	Public Switched Network
QPSK	Quaternary Phase Shift Keying
REI	Radio Exchange Interface
RMS	Root-Mean-Square
RPE-LTP	Regular Pulse Excitation-Long-Term Prediction
SA	Simulated Annealing
SAE	System Architecture Evolution
SAT	Supervision Audio Tone
SC	Setup Channel
SISO	Single-Input Single-Output
SDMA	Space Division Multiple Access
SINR	Signal-to-Interference plus Noise Ratio

SIR	Signal-to-Interference Ration
SNR	Signal-to-Noise Ratio
SON	Self-Organizing Network
TACS	Total Access Communication System
TDD	Time-Division Duplex
TIA	Telecommunications Industry Association
TIREM	Terrain Integrated Rough Earth Model
TDMA	Time Division Multiple Access
UHF	Ultra High Frequency
UMTS	Universal Mobile Telecommunications System
VC	Voice Channel
VHF	Very High Frequency
VIS	Vector Immune System
VSELP	Vector Sum Excited Linear Predictive coding
W-CDMA	Wide-Band Code Division Multiple Access
WSS	Wide Sense Stationary
WSSUS	Wide Sense Stationary Uncorrelated Scattering

1

Mobile Cellular Telephony

1.1 Introduction

A mobile telephone system is defined as a communications network via radio which allows continuous mobility by dividing its coverage area into cells. Wireless communications, on the other hand, implies radio communications without necessarily requiring handover from one cell to another during an ongoing call (Nanda and Goodman, 1992).

The cellular communications network was originally proposed by Douglas H. Ring, who worked for the Bell Telephone Laboratories, in the USA, in 1947. The paper was an internal memorandum, and it was never published outside the laboratory. The abstract of the memorandum read (Ring, 1947):

> "In this memorandum it is postulated that an adequate mobile radio system should provide service to any equipped vehicle at any point in the whole country. Some of the features resulting from this conception of the problem are discussed along with reference to a rather obvious plan for providing such service. The plan which is outlined briefly is not proposed as the best solution resulting from exhaustive study, but rather is presented as a point of departure for discussion and comparison of alternative suggestions which may be made."

Experiments with the new mobile cellular technology started, in 1978. The first country to offer a cellular service was Switzerland, in 1981, although Motorola had patented the cellular telephone in 1973. In spite of its release in 1979, in Chicago, the system only started full installation, in 1984, in the USA.

The effort of the industry to provide ever more efficient mobile communications via radio to the population, has demanded intense research and development over the years (Hashemi, 1991; Dhir, 2004). One of the aims

1

of this research is to avoid high costs associated with the installation and relocation in places interconnected by wires (Freeburg, 1991).

Various cellular telephone systems have been deployed and some are operational to handle the control and the information flow in mobile systems. The main systems are the following:

1. Frequency (FM) division multiple access (FDMA): the first access technology, that uses different frequencies to transmit the signals.
2. Time division multiple access (TDMA): in this access system the signals are allocated in separate time slots.
3. Code division multiple access (CDMA): in which the access is provided by a set of orthogonal codes, designed to minimize the correlation between the transmitted signals.
4. Space division multiple access (SDMA): the access in this system is obtained by the use of an antenna array, which separates the signals in distinct beams.
5. Multiple-input and multiple-output (MIMO) is a method to improve the transmission rate of a radio link using multiple transmit and receive antennas to exploit properties of the multipath propagation.

A conventional mobile telephone system selects one or more radiofrequency (RF) channels for use in specific geographic areas. The area of coverage is planned to be as large as possible, which may require a rather large transmitting power.

In a cellular mobile telephone system, the area of coverage is divided into regions called cells, in a manner that keeps the transmitted power as low as possible, and permits the reuse of the available frequencies.

1.2 First Systems to Enter Commercial Operation

The mobile cellular analog telephone systems that first entered commercial operation were very similar. All of them, in one way or another, were based on the advanced mobile phone service (AMPS), developed in the USA in the seventies.

The procedures for completing and maintaining a call are similar to those of the digital systems, but are easier to understand using the analog procedures. For this reason the analog system will be used as a reference in this chapter.

In the AMPS system, the voice signals modulate the channel carriers in FM. Signaling, which operates at a 10 kbit/s rate, uses frequency shift keying

(FSK), that is, digital FM. Voice channels, as well as, signaling channels occupy, individually, a bandwidth of 30 kHz.

The use of FM modulation, that is, of a constant envelope carrier, in the AMPS standard has the purpose of minimizing multipath effects. The multipath effect, common in the mobile channel, produces a multiplicative noise which acts on the carrier as an amplitude modulating signal.

For amplitude modulated signals, the multiplicative noise would be incorporated into the modulating signal, making it difficult to recover the original modulating signal. Angle modulation (FM or phase) can eliminate part of the noise by limiting (clamping) the carrier amplitude before demodulation. This does not affect the signal recovery, since the desired information is contained in the FM, or phase, of the carrier.

An important characteristic of these systems is the protection techniques used in the transmitted messages. Some of the analog systems in the world were: NTT (Japan), AMPS (USA and Brazil), TACS (UK), NMT (Scandinavia), and C-450 (Germany), which used the following protection methods:

- Principle of majority decision (PMD) – in which various copies of the signal are transmitted.
- Automatic repeat request (ARQ) – in which whenever an error is detected, then a signal retransmission is requested.

In environments with strong fading, the PMD is a good option, but environments with slow fading require ARQ. The AMPS system uses PMD. Table 1.1 summarizes the main technical characteristics of the main analog standards, in which it is noted that the AMPS and TACS standards have 42 control channels.

The analog system were used in many countries for interconnecting different digital standards, and constituted an important basis for mobile cellular telephony, and frequency modulation was used in all deployed systems. The basic principles of digital telephony are independent of the specific type of system employed.

1.3 Description of the Cellular System

The design of a cellular system consists in dividing the region to be covered by mobile telephony into smaller areas, to allow the use of low power transmission and efficient use of the spectrum by means of frequency reuse.

Table 1.1   Technical characteristics of analog mobile communications standards

Characteristics	Japan	USA	UK	Germany
System	NTT	AMPS	TACS	C-450
FM band (MHz)				
Base to mobile	870–885	869–894	935–960	461.3–465.74
Mobile to base	925–940	824–849	890–915	451.1–455.74
Frequency spacing between Tx and Rx (MHz)	55	45	45	10
Channel spacing (kHz)	25	30	25	20
Number of channels	600	832	1000	122
Coverage radius (km)	5–10	2–20	2–20	5–30
Audio signal				
Modulation type	FM	FM	FM	FM
Frequency deviation	± 5	± 12	± 9.5	± 4
Control signal				
Modulation type	FSK	FSK	FSK	FSK
Frequency deviation	± 4.5	± 8	± 6.4	± 2.5
Data transmission Rate (kbit/s)	0.3	10	8	5.28
Message protection	ARQ	PMD	PMD	ARQ

Therefore, cellular systems are limited by interference, produced by the other users, while conventional mobile systems, that use high-power levels, is limited by thermal noise.

1.3.1 Cellular Structure

In principle a given region or coverage geographic area to be served by the mobile cellular service is divided into sub-regions, called cells. The cell is a geographic area *illuminated* by a base station (BS), within which signal reception follow certain system specifications.

Extension of the coverage area of a BS depends on the following:

- Radio transmitter output power, which has to be limited to minimize interference.
- Frequency band employed, because the attenuation produced by the atmosphere, by the city topology and structure, are dependent on the carrier frequency.
- Antenna height and location, which determines if the mobile station has a clear view of the transmitting antenna.
- Antenna type, which defines the gain and directivity.

- Terrain topography is important to compute the transmission losses.
- Receiver sensitivity establishes the minimum power level that can be detected.

In conventional systems, the approach used to cover a large area is to radiate a relatively high power level. This technique, however, is not appropriate for cellular systems, except in areas with a very low traffic density, such as rural areas, or with coverage by large cells.

Electromagnetic waves propagate from a BS in a straight line, known as line-of-sight. This means that a user connected to a BS located behind a large obstacle, such as a hill or a tunnel, can be in an area without radio coverage, also called an area of shade.

However, the presence of buildings in large cities is not a critical impediment to propagation of radio waves because of their reflection and refraction properties, which minimize the shade effect. On the other hand, areas with a high traffic density are usually divided into a large number of small cells.

The output power at the BS transmitter can be adjusted, so the coverage area can be increased or reduced to the appropriate dimension. Currently, there are cells with a coverage radius of less than 500 m for outdoor environments and micro-cells, with just a few meters of radius, that are used in shopping centers, viaducts, and convention centers.

For the initial design, the operator of the cellular system divides an area into equal size cells of hexagonal shape, in a manner that the center of each cell contains a BS. This type of cell is graphically represented by a hexagon, because it appropriately tessellate the plane. The most common types of cells are the following:

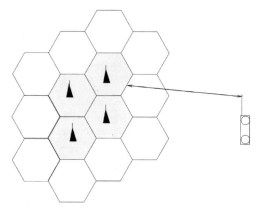

Figure 1.1 Description of the cellular system.

- **Omnidirectional cell** – in omnidirectional cells, the BS is equipped with an omnidirectional antenna, that is, an antenna which transmits the same power in all directions, in the azimuthal plane, forming thus a coverage area approximately circular, with the BS in the center.
- **Sector cell** – the BS in each sector cell has a directive antenna, to cover a specific region.

1.3.2 Cellular Structures

A group of neighboring cells which use the whole group of 395 voice channels and 21 control channels available in the advanced mobile phone system (AMPS) is called a cluster, or set, of cells. In other words, within a cluster there is no frequency reuse.

The number of cells forming cluster differs according to the cell structure employed. The most common cell structures are the following:

- Four cell standard, all omnidirectional cells, as shown in Figure 1.2;
- Seven cell standard, all omnidirectional cells, illustrated in Figure 1.3;
- Twelve cell standard, all omnidirectional cells.
- Twenty-one cell standard, with seven BSs, each BS associated to three sector cells.
- Twenty-four cell standard, with four BSs, each BS associated to three sector cells.

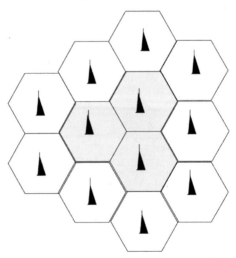

Figure 1.2   Sets of omnidirectional cells – three cells $(i = 1; j = 1)$.

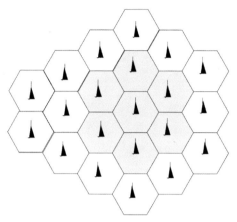

Figure 1.3 Sets of omnidirectional cells – seven cells ($i = 2; j = 1$).

A cluster with 21 cells has each pair of channels spaced by 21 other channels. This characteristic, facilitates transmitter use, and also minimizes the risk of possible adjacent channel interference within a cell. The channel allocation planning takes into consideration the traffic density between various cells.

1.4 Frequency Reuse

Frequency reuse means the simultaneous use of the same frequency in two distinct sets of cells. The distance between cells for frequency reuse (reuse distance) is limited by the maximum co-channel interference allowed in the system.

In the case of a homogeneous system, formed by sets of

$$N = i^2 + ij + j^2 \tag{1.1}$$

cells, in which i and j are positive integers, the reuse distance D is given by

$$\frac{D}{R} = \sqrt{3N}, \tag{1.2}$$

in which R is the cell, or coverage, radius.

Evidently, the quality of the transmissions is reduced as the size of the clusters becomes smaller, but the traffic capacity increases due to the possibility of distributing all channels among a few cells. Table 1.2 shows the existing relationship between traffic capacity and co-channel interference, for various hypothetical sizes of cell sets (Yacoub, 1993a).

Table 1.2   Traffic capacity and co-channel interference

Dimension	D/R	Channels/Cell	Traffic Capacity	Transmission Quality
1	1.73	360	Higher	Lower
3	3.00	120		
4	3.46	90		
7	4.58	51		
12	6.00	30	Lower	Higher

Sets of frequencies, with no frequency in common, can be allocated to neighboring cells, since a total radio coverage of a certain region requires superposition of cells. However, if a particular frequency is used in neighboring cells, the so called co-channel interference is likely to appear in the superposition areas.

Interference is the factor that pushes for a substantial increase in the separation distance between two cells using the same frequency. The repetition distanced and the reuse of the same frequency in different cells is called frequency reuse. Due to the frequency reuse technique, the maximum number of voice channels in the AMPS system was a multiple of 395, in the so-called extended AMPS.

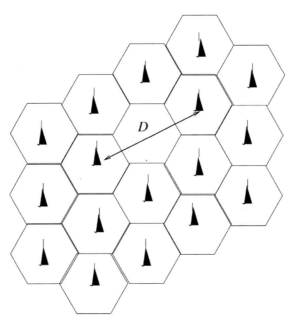

Figure 1.4   Concept of mean reuse distance.

1.4.1 Cell Division

Consider that a new cellular system has been deployed, which covers a large city and its neighboring region. Given that, in the near future, the subscriber traffic density will be low, then there will be no need to expand the system immediately. Then, there is no need to provide an excessive capacity at a high cost, and without a quick return for the investment.

As a first step, in a green field scenario, a BS could be installed, with a high antenna and an output power sufficient to cover a radius of 20 km, for example. Considering a cluster structure, the BS could allocate a group of frequencies like, for example, group A, and consequently the BS would be limited to approximately 45 voice channels and three control channels.

A possible way to increase capacity within a given coverage area is to place, for example, three more superimposed BSs. The new sites could temporarily work as omnidirectional cells, but in future they can be replaced by sector cells. The frequency group must be defined according to the frequencies that the new BSs will use in the future.

As time passes it may become necessary to increase the capacity in the central area. For that purpose, the coverage of the BS in a group is reduced, by means of a reduction in its power level, which causes a reduction of the coverage area. In spite of the cells designed to cover suburbs, they can in some case provide an auxiliary coverage to the central area.

When the demand for traffic grows within a given cell, there are two possible solutions to handle more traffic: addition of new cells or the sectoring of existing cells.

- Addition of new cells – The power of the transmitters in the existing cells is reduced enough to cover, for instance, half of the original coverage area and new cells are added to meet the additional need for coverage. The addition of new cells implies in the installation of towers and antennae, with their respective costs.
- Cell sectoring – The set of omnidirectional antennas in a cell is replaced by directional antennas (with 60° or 120° aperture) and the cell is divided into sectors. Cell sectoring is a more economical solution because it uses exiting structures.

The cell size must be adequate to the telephone traffic density. The higher the traffic the smaller the corresponding cell size, since the number of voice channels available per cell is limited. This implies, for example, that in areas near the city center the cells are smaller than those in suburban areas.

The addition of new cells is a more flexible manner to expand the system, but it is also a most expensive one. It involves the installation of new BSs and antennas. The acquisition of land for the installation of BSs in city central areas became prohibitive for the operators, which prefer to rent a small area on the top of buildings or other commercial areas. Cell sectoring has a smaller cost for the operator, however it imposes restrictions to future expansions.

The subdivision of large cells into small cells implies that the frequency reuse distance becomes smaller and that the number of channels within the same geographic area is increased, thus increasing system capacity.

The technique of super-imposing cells is employed in different situations to solve specific problems. The super-imposed cell receives calls from several BSs. The border region between large and small cells provides a typical example.

It is possible to define a super-imposed cell as a group of voice channels allocated in the same region of a common cell, which receives the name of sub-imposed cell. A super-imposed cell differs from a sub-imposed cell because the latter not only has a smaller radius, but does not have a signal intensity receiver and a control channel.

1.5 Management of Channel Utilization

The spectrum allocation in the AMPS standard considered separate frequencies for the transmission from the BS and transmission from the mobile terminal (MT; full duplex operation). The difference between the two frequencies was 45 MHz.

The AMPS system operated in the frequency band of 825–845 MHz (824–849 MHz in the extended system) for transmission from the MT to the BS, and the frequency band of 870–890 MHz (869–894 MHz in the extended system) for transmission in the reverse direction. Most of the channels (having 30 kHz bandwidth) is allocated for telephonic conversation.

The remaining channels transmit signaling information in digital form. The channels used for signaling are called setup and are used for exchanging messages required to establish a call.

1.5.1 Channel Allocation in the AMPS System

In the AMPS system, the channel frequency allocation was done as follows, in which the figures in parenthesis refer to the extended system.

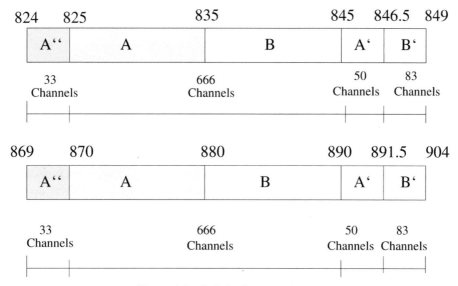

Figure 1.5 Cellular frequency bands.

- 825–845 MHz: 666 channels with bandwidth 30 kHz each (824–849 MHz: 832 channels with bandwidth 30 kHz each) for communication from the MT to the BS;
- 870–890 MHz: 666 channels with bandwidth 30 kHz each (869–894 MHz: 832 channels with bandwidth 30 kHz each) for communication from the BS to the MT;
- The set of 666 (832) channels is divided into two independent systems, A and B, such that each can be exploited by a different operator;
- Each operator has available 333 (416) channels and can market 312 voice channels, because the remaining 21 channels are required for control purposes;
- In the original standard, channels are numbered from 1 to 333 in system A, and from 334 to 666 in system B, beginning from the lowest channel frequency available in each system.

A channel center frequency can be calculated from its corresponding channel number as follows:

- Direction from BS to MT: F = (channel number $\times$ 0.03) + 869;
- Direction from MT to BS: F = (channel number $\times$ 0.03) + 824.

1.6 Constitution of the Cellular System

A typical cellular system consists of three elements, including the connections between them. The basic components of a cellular system are as follows (CNTr, 1992):

- Mobile terminal (MT).
- Base station (BS).
- Mobile Switching Center (MSC).

The MT provides the air interface for the user of the cellular system, contains also a control unit, a transceiver and an antenna. The MT transmits and receives voice signals, thus making conversation possible, and transmits and receives control signals, thus allowing the establishment of a call.

Mobile terminals are produced by a large number of manufacturers, which is the reason for a wide variety of designs and facilities for the subscribers. The MT's can be used in a variety of applications, such as:

- Installed in a car (vehicular cellular telephone), the first devices in the early times.
- Portable (portable cellular telephone), illustrated in Figure 1.7.
- Installed in a rural area (rural cellular telephone). illustrated in Figure 1.8.

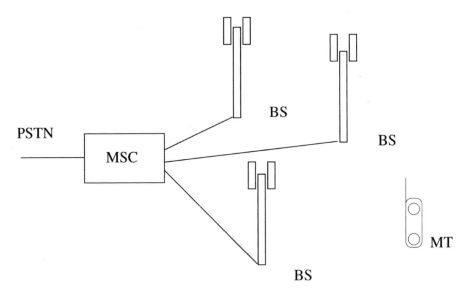

Figure 1.6 Components of a cellular system.

Figure 1.7   Example of a portable cellular telephone (Courtesy of Nokia).

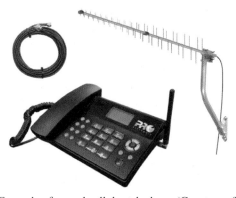

Figure 1.8   Example of a rural cellular telephone (Courtesy of Quadriband).

The maximum output power of a unit installed in a car or in a rural area is 3 W, while the power of a portable unit is 0.6 W for American standards, and 1 W for the European standard. When an MT assesses an MSC, it sends its station class in which is indicated its maximum output power. An MSC keeps a file with parameters of all registered cellular terminals, for call control purposes.

In order to be identified by the system, each MT has a mobile identification number (MIN) and, for security reasons, each mobile unit has an electronic serial number (ESN), which is defined by the manufacturer.

The signal quality monitoring which reaches the MTs is performed by the audio monitoring tone, or supervision audio tone (SAT), and by the

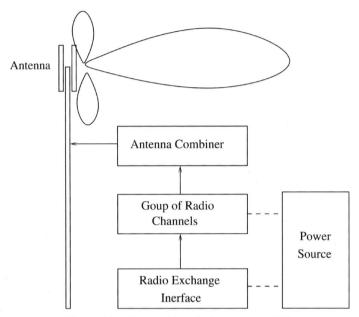

Figure 1.9   Composition of a base station (BS).

measurement of the intensity level of signals received by the MTs, in the analog case, and by the value of the bit error rate (BER) in digital systems.

A BS encompasses the following functional units, which are illustrated in Figure 1.9

- Group of radio channels (GRCs);
- Radio exchange interface (REI);
- Antenna combiner; Antenna, which is illustrated in Figure 1.11; and
- Power source.

The GRCs contains voice channels and control channels. The radio exchange interface (REI) operates with an adapter for voice signals between the MSC and the BS, since the coding for the aerial interface is different from the coding used for communication between telephone exchanges.

Therefore, the equipment receives data and voice from the channel units and sends information to the MSC by means of a dedicated link from BS to MSC. In the opposite direction the equipment receives data and voice from the MSC by means of a link from MSC to BS, and sends them to the corresponding channel unit or control unit. Among other things a BS takes care of the following:

Figure 1.10 Example of a cellular BS, (Courtesy of Siemens, Deutches Museum).

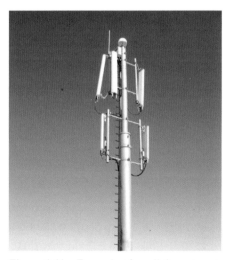

Figure 1.11 Example of a cellular antenna.

Provides the interface between an exchange and control unit and MTs;

- Contains a control unit, radio transceivers, antennas, power plants and data terminals;

- Transmits and receives control signals for establishing call monitoring;
- Transmits and receives voice signals from various MTs within its coverage area.

In the connections between BSs and MSC's, two types of facilities are employed.

- Trunks to provide a voice communication path. The number of trunks is decided based on the traffic and on the blocking probability desired (desired grade of service);
- Trunks to provide a signaling path to the establishment and monitoring of calls.

Since a BS transmits and receives control signals for the establishment and monitoring of calls, as well as voice signals from various MTs within its coverage area, it can be said that the BS acts as a traffic concentrator for the MSC.

The switching and control center, also called MSC, is considered the main element of the mobile cellular system. Among its functions it is worth mentioning the following:

Central coordination of the whole cellular network, managing all BSs within its control area, that is, switching and controlling the cellular aggregate;

- It works as an interface with the Public Switched Network (PSN);
- It switches calls originated or terminated in a mobile station;
- It allows the mobile station to have available the same services provided by the public network for fix subscribers; and
- The MSC connects to the BSs via optical cable. Sometimes, the connection uses a microwave link.

1.7 Characteristic Functions of a Cellular Network

A complete cellular system may contain many MSC's, which can be seen as functional interfaces to the switched public telephone network, and the signaling employed for establishing calls will depend on that signaling defined by the switched public telephone network.

On the other hand, each mobile station is connected by software to a unique MSC, which is normally the one in the subscriber home area. This MSC is called the home center office (MSC-H).

The MSC operates on a trunk-to-trunk basis, and has three types of interconnecting links, namely:

• Interconnects the MSC to a local public switching telephone exchange.
• Interconnects the MSC to a tandem public switching telephone exchange.
• Interconnects the MSC to a local exchange together with type a tandem exchange, over a high usage alternative routing basis.

1.7.1 Handoff

Handoff, or handover, is a mobile function that allows the continuity of a conversation to be preserved when a user crosses from one cell to another. The handoff is centralized in the MSC and in the AMPS system it causes a communication interruption of less than 0.5 s.

The Global System for Mobile Communications (GSM), or Groupe Spécial Mobile, as it used to be called by the European Telecommunications Standards Institute (ETSI), has four types of handoff:

1. Intra-BTS handoff – This happens when, for instance, it is required to change the frequency band, or time slot, being used by a mobile because of interference. The mobile remains connected to the same BS transceiver, but changes the channel or time slot.
2. Inter-BTS and Intra BSC handoff – This occurs when the mobile moves out of the coverage area of a base transceiver station (BTS), but inside another one controlled by the same BS controller (BSC). In this case the BSC executes the handoff and assigns a new channel and slot to the mobile.
3. Inter-BSC handoff – When the mobile moves out of the range of the cells that are controlled by one BSC, a more complex handoff is performed. It changes from a BSC to another, which is controlled by the MSC.
4. Inter-MSC handoff – This occurs when the mobile changes between networks. In this case, the two MSC's involved negotiate the control of the process.

1.7.2 Roaming

When a mobile station enters the control area of an area which is not related to a home MSC, this switching center is known as the visited center office

(MSC-V), and the subscriber is called the visitor. Calls for a subscriber in this condition are routed and switched by the MSC-V.

The concept of a mobile station moving from one control area to another is called roaming. Signaling between MSC's can be implemented, for example, according to a protocol called common channel signaling number 7 (CCS-7) of the ITU-T, by means of a direct link connecting two MSC's or by a switched public telephone network (Alencar, 2011).

The following procedures are employed when the roaming function is used (Alencar and da Rocha, 2005):

1. When a mobile station enters a new control area it is automatically registered in the MSC that controls that area.
2. The visited MSC checks whether the mobile station had already been registered earlier. If that was not the case, the visited MSC will inform the home MSC of this new condition.
3. The home MSC registers the service area being visited by the subscriber. After this procedure is completed, the visiting subscriber is then able to originate and to receive calls, as if the user was in the home area.

The change from a BS connected to an MSC to another BS connected to another MSC, during a call in progress, is called *call handoff* between control centers. In some systems, the control areas can still be subdivided into location areas.

When a MT passes from one location area to another, it must inform the MSC about its new location. This task is known as forced registration or as registration of the location area. In this case the search for the MT is done only in those cells that belong in the location area.

The forced registration implies updating the MSC files the at predefined time intervals. MTs perform a routine of generation of a random time value (time instant), starting from a seed sent by the BS, for sending location information.

The maximum period for sending data is defined by the control center operator. The periodic updating of registers might overload the central processor and degrade the MSC Grade of Service (GoS), and has to be carefully planned (Alencar, 2011).

The registration per location area is done every time a subscriber crosses the limit pre-established by the system operator. This limit can be, for example, the cell or the coverage area of the MSC. The larger the area to be covered, for locating a subscriber, the larger the amount of allocated resources of the mobile cellular system.

This procedure can block channels and also degrade the GoS of the system. A combination of the two techniques is, usually, more efficient, and several algorithms have been developed to attain this objective.

1.7.3 Configuration of the Mobile Cellular System

A BS, or, sometimes, a BS controller, is connected to an MSC by means of point-to-point circuits. The BS has as its main function to work as a repeater for voice information and data, as well as to monitor the transmission link quality during a conversation.

The connections between a BS and a MSC are done by radio, in the microwave range, or by optical fiber. The MSC's are, usually, connected to the public switched telephone network via optical fibers.

1.7.4 Data Communication from the MSC to the BS

In a cellular system there is a constant exchange of information between the control central and the BS. The communication between the MSC and the BS occurs in the following cases (Alencar and da Rocha, 2005):

- Whenever a user of the MSC originates a message for a mobile station, to be sent through the control channel or through the voice channel.
- Whenever the MSC receives a message from a mobile station, requiring a call setup, for example.
- Whenever the MSC receives a request from the BS, or BS controller, such as, for example, a handoff request.
- Whenever the MSC sends a message to the BS such as, for example, requests for the values of measurements made in a handoff process.
- Whenever the MSC receives an alarm message due to failure in the radio link such as, for example, a transmitter problem or a power failure.
- Whenever the MSC receives a message originated from a BS, such as, for example, a power level measurement.
- Whenever an external alarm signal is detected, for example, due to the presence of an intruder in the BS or due to a fire.
- Whenever maintenance routines must be executed such as, for example, loading channel units from a memory bank in a MSC or a routine test for these units.

1.8 Radio Channel Types

In a cellular mobile system, there are two main types of transmission channels between a BS and a MT. The types are:

1. Voice channels (VC) – During the establishment of a call the MSC selects a voice channel, which will transport the conversation. When the conversation is finished, this voice channel is free and subsequently becomes part of a file managed by the MSC, that is updated with various channels and their respective states.

Whenever a voice channel is free the transmitter in the channel unit at the BS is switched off, being switched on again only when the that channel is again captured. These actions are all controlled by the MSC.

In order to control the establishment of a call the MSC needs to exchange information constantly with the MT, by means of the voice channel. Because of that, other signals, besides voice signals, flow through this channel.

1. Control channels, also known as setup channels (SCs) – The control channel is the means for exchanging signaling in the cellular system air interface. It is responsible for allowing a conversation to be established. Usually there is only one control channel in each cell. In general, the control channel is used for the following tasks:

    a. To transmit general purpose messages with system parameters.
    b. To transmit messages for search and general information.
    c. To transmit messages for designating a voice channel.
    d. To receive messages from the MT.

The MT, when moving from one cell to another, stops tuning in the control channel from the original cell and synchronizes the control channel of the new cell. This procedure characterizes the o handoff.

There are two types of control channels:

1. Forward control channel, and
2. Reverse Control Channel.

The messages sent through the forward control channel are as follows:

1. Paging for a MT;
2. Continuous transmission of messages to the MTs;
3. Designation of a voice channel;
4. Directed trial, which indicates an adjacent cell to which the call must be directed; and
5. Overhead, which allows the MTs to adjust to various systems, including different manufacturers.

The connection between a MT and a BS (upstream uplink) is done eventually for sending information. Messages are transmitted when a MT tuned to the control channel has meaningful data to send, as follows:

1. Access of a MT to originate a call;
2. Confirmation of orders; and
3. Answer to the paging signal.

1.9 Second Generation Systems

The development of digital systems, which constitute the basis for the second generation of mobile cellular systems, followed different paths in distinct regions of the World. In the USA, the main concern was to maintain compatibility with the AMPS system and to allow roaming in the country.

In Europe, since there were many diverse systems, the objective was to develop a standard compatible with the fixed telephone network and with International Telecommunication Union (ITU) standards. The main objectives of a digital mobile communication system are:

- Improve the spectral efficiency;
- Reject interference as much as possible;
- Immunity to the propagation environment;
- Increase the receiver sensitivity; and
- Use error-correcting codes to protect voice signals.

By considering the previous objectives, the goals established for the transition between the analog standard and the digital cellular system, as defined by the Federal Communications Commission (FCC), in USA, were as follows (Alencar and da Rocha, 2005):

- Accommodate more subscribers within the same frequency band, with the use of a time division multiple access (TDMA) or code multiple division access (CDMA).
- Provide adequate capacity to expand the coverage for the subscribers.
- Guarantee performance quality in the transmissions of the subscribers.
- To minimize the impact hardware and other costs.

In Europe, as a consequence of the multiplicity of existing analog systems, from the beginning there was no compatibility requirement. The digital GSM technology made international roaming possible, which was not the case earlier due to the various diverse existing standards.

The USA and Japan chose to keep a certain degree of compatibility with their first generation systems, to allow, by means of dual telephone

Table 1.3   Comparison among various digital cellular systems

System	Europe	USA	USA	Japan
Parameter	GSM	IS-54	IS-95	Personal digital cellular (PDC)
Band (MHz)				
Base → Mobile	935–960	869–894	869–894	810–826
Mobile → Base	890–915	824–849	824–849	940–956
Multiple access	TDMA	TDMA	CDMA	TDMA
Channels/carrier	8	3	55/62	3
Voice encoder	Regular pulse excitation-long-term prediction (RPE)-LTP	Vector sum excited linear predictive coding (VSELP)	CELP	VSELP
Channel rate uncoded (kbit/s)	13.0	7.95	1.2/9.6	6.7
Channel rate coded (kbit/s)	22.8	13.0	19.2/28.8	11.2
Modulation	GMSK	$\pi/4-$ DQPSK	QPSK/DPSK	$\pi/4-$DQPSK
Transmission rate (kbit/s)	270.8	48.6	1228.0	42.0
Min. cluster	4 (pessim.) 3 (optim.)	7 (pessim.) 4 (optim.)	1 1	7 (pessim.) 4 (optim.)
Adaptive equalizer	Mandatory	Mandatory	No	Optional
Bandpass (in kHz)	200	30	1250	25

sets, a soft transition into the digital technology. However, as can be seen in Table 1.3, such systems have some points in common, although with distinct ways of implementation (Falconer et al., 1995). For example, all systems use hybrid source encoders, benefiting from the combination of high quality waveform encoders and the high compression efficiency of parametric encoders (vocoders; Steele, 1993).

The operation of high efficiency non-linear power amplifiers, required variations of the four-level modulation by phase deviation quaternary phase shift keying (QPSK), which also provided a satisfactory compromise between the transmission channel bandwidth and tolerance to noise. The TDMA access technique is used in one of the American standards and in the European and

Japanese standards. The other American standard adopted the CDMA access technique (Assis, 1994).

1.9.1 The European GSM Standard

The development of a cellular mobile telephone system for the European countries, employing digital technology, began in 1982, with the Conférence Européene des Administrations des Postes et Télécommunications (DEPT), in a work group called Groupe Spéciale Mobile (GSM; Alencar and da Rocha, 2005).

The basic specification of the European digital standard, or GSM standard as it became known, including the TDMA access technique and the system architecture, was approved, in 1987. Later, GSM was adopted by the European Telecommunications Standards Institute (ERSI) created, in 1988. The field tests with GSM began in July 1991, and its introduction as a commercial service in Europe started in 1992. After the GSM standard GSM was concluded, the acronym meaning was changed to Global System for Mobile Communications.

This standard was adopted in all of Europe, in Australia, in Brazil, and in various countries in Asia and Africa. The GSM remained the principal world standard for more than two decades.

It is noticed in Table 1.2 that the bandwidth per carrier is 200 kHz. In the TDMA access technique, each user has all of this bandwidth available for 577 μs, that is, for the period of a user time slot. In GSM a frame consists of a group of eight distinct users, the duration of which is 4.615 ms (8 × 0.577 ms). This structure simplifies the RF part of the BSs and of the MTs, since an independent receiver for each user is not required.

The handoff process has also a characteristic that is distinct from that of the analog systems. It is noted that the MT uses only 1.18 ms of a total 4.615 ms of frame duration to transmit and to receive messages from the BS. The time duration (480 ms in case of GSM) and the six strongest levels are sent via a control channel to the BS which is connected to the MT. This procedure is known as *handoff* with the help of the MT mobile-assisted handoff (MAHO).

The Gaussian minimum shift keying (GMSK) modulation is derived from the minimum shift keying modulation technique. In GMSK the modulating signal is passed through a Gaussian filter before entering the modulator. The advantage of this modulation technique comes from the fact that it has a constant envelope, which permits the use of power amplifiers operating in the non-linear region, in which their efficiency is higher.

The transmission rate at the output of the source encoder is 13 kbit/s regular pulse excitation-long-term prediction (RPE-LTP). By adding 9.8 kbit/s of the error correction circuitry (rate 1/2 convolutional code) and 11.05 kbit/s for the control channel and synchronism, a total rate of 33.85 kbit/s per user results and an overall rate of 270.8 kbit/s for a frame.

To compensate for the fading effects due to multipath propagation, the GSM standard employs channel equalizers. The convolutional code, with bit interleaving and frequency hopping (meaning that each user time slot can be transmitted with a distinct carrier frequency, if this feature is acquired from the vendor by the operator), are the techniques employed to improve GSM performance. Frequency hopping is an optional function in the BS but it is mandatory in the MT.

The GSM standard was designed so as to be compatible with the integrated services digital network (ISDN), defined by the ITU. Consequently it offers various call options as, for example, wait, call redirecting, calls with restriction plus a variety of data services up to 9600 kbit/s, and specific modems are not required (Assis, 1994).

1.9.2 Second Generation American Standards

The need to develop a system with higher capacity than that obtained with the AMPS standard became evident in USA, in 1988, with the overload of the cellular mobile service in large cities. On the other hand, adoption of a completely new standard could bring problems to millions of users of the AMPS system, besides requiring a high cost infrastructure for its implementation.

Therefore, the Telecommunications Industry Association (TIA) decided to implement a system which allowed an increase in service capacity and was compatible with the AMPS system.

1.9.2.1 The TDMA standard

Requiring compatibility with the existing system, the TDMA standard using 30 kHz of bandwidth per carrier frequency was approved by the TIA. Originally this standard was designated as IS-54 (Interim Standard). Later it became known as D-AMPS, or digital AMPS. This standard employs a TDMA with three users per carrier. The frame has a duration of 20 ms, with each channel window corresponding to 6.67 ms.

The voice encoder output (VSELP – vector sum excited linear predictive coding) transmission rate is 7.95 kbit/s. The error correcting circuit (rate 1/2 convolutional code) contributes 5.05 kbit/s. There is still 3.2 kbit/s due to the control channel and the e synchronization bits. In this manner, the rate per user is 16.2 kbit/s, which leads to a total transmission rate of 48.6 kbit/s. At this rate the use of equalizers could be optional, and is recommended only in situations in which the delay exceeds 2 μs. However, the use of equalization is specified in the D-AMPS standard.

The $\pi/4$ – QPSK modulation limits the carrier phase transition to $\pm\pi/4$ and $\pm3\pi/4$. Because there are no $\pm\pi$ transitions as in a conventional QPSK modulator, the envelope fluctuation is substantially reduced, which allows the use of non-linear power amplifiers. Differential encoding is employed together with a filter with a 0.35 roll-off factor.

The handoff process is similar to that described for the GSM. The only difference is that in the D-AMPS the measurements are performed in at most 12 BSs, and the average of these measurements is taken over a period of 1 s. The result of all measurements is sent to the BS in use.

1.9.2.2 The CDMA standard

The use of spread spectrum as an access technique in mobile cellular systems is a clever way to reject interference, and take advantage of the multipath signals. The use of orthogonal codes and a careful control of the power levels transmitted by a BS are fundamental procedures. As a consequence of this property, the following characteristics are important (Lee, 1991).

1. Reuse of all available frequencies – since all cells share the same RF channel, frequency coordination becomes unnecessary.
2. Use the multipath signals to improve the reception, avoiding the use of channel equalizers – in a multichannel receiver (called RAKE) used in the CDMA system, demodulation of each signal arriving by a different path (multipaths) is done individually, thus obtaining an efficient effect of diversity.
3. Soft handoff procedure – the various channels in the RAKE receiver gradually replace the code of the BS in operation by the corresponding code for the new BS. The handoff is not abrupt, as in systems that employ FDMA or TDMA techniques.
4. It naturally allows the use of the vocal activity factor to increase system capacity – considering that during a conversation on average the channel is activated only 40% of the time, since the lack of vocal activity implies in no signal present in the system for transmission. This characteristic is exploited to increase system traffic capacity.
5. Ease of sharing frequency with other systems – when under interference, the CDMA system offers high rejection to interfering signals. As a source

of interference, within certain limits, the perturbation introduced by the CDMA system can be controlled, since the spectrum spreading produces a relatively low power density in the used frequency band.

However, apart from using a more complex technology there are two concerns related to the use of spread spectrum for CDMA:

1. Near-far interference – since all MTs share the same channel, a strong signal from a user near the BS can mask a weak signal from a user far from the BS. There is need for an effective control of the power level in the MT. In practice this control covers a 100-dB range.
2. Need for synchronization – a synchronization failure can harm the system operation. In order to avoid such a situation occurring, the BSs are synchronized by means of signals coming form a global positioning system (GPS) satellite, whereas MTs receive synchronization signals from the BSs.

The CDMA system was recognized by the TIA, in 1992, which conferred to it the designation IS-95, becoming then the second digital American standard. The system commercial name is *cdmaOne*.

1.9.2.3 cdmaOne

The *cdmaOne* is considered the second generation (2G) of mobile wireless technology. The system was supported by the CDMA development group (CDG) and is compatible with the air interface IS-95 and with the network standard for exchange interconnection ANSI-41. The protocol, version IS-95A, operates in the frequency bands of 800 MHz and 1.9 GHz, and employs a data transmission of up to 14.4 kbit/s. Version IS-95B supports up to 115 kbit/s, by employing eight channels.

1.9.2.4 The Japanese standard

This standard personal digital cellular (PDC) is similar to the described American TDMA system, but operates in the 800 and 1.500 MHz frequency bands. A few differences such as frequency bands and transmission rate of the voice channel can be observed in Table 1.3. One distinct aspect in this standard is that the use of an equalizer is optional (Assis, 1994).

1.10 Third Generation Systems

A few systems were considered for the third generation of mobile cellular telephony, which included the ITU Universal Mobile Telecommunication

System (UMTS), the General Packet Radio Service (GPRS), an evolution from GSM, and the North American cdma2000 standard.

1.10.1 GPRS

The General Packet Radio Service (GPRS) was developed for data transmission in the GSM system. It uses the network only when there are packets ready for transmission, at a rate of up to 115 kbit/s.

1.10.2 UMTS

The Universal Mobile Telephone Standard (UMTS) proposes a data transmission rate of up to 2 Mbit/s by means of a combination of TDMA and W-CDMA (Wide-Band CDMA), operating in the 2 GHz band. UMTS is the European member of the IMT-2000 family of third generation (3G) cellular systems. The data rates offered to users are as follows: 144 kbit/s for cars, 384 kbit/s for pedestrians, and 2 Mbit/s for stationary users (Dhir, 2004).

1.10.3 cdma2000

The cdma2000 standard offered users of the IS-95 standard a soft transition to the 3G cellular system. This standard is also by the designation ITU, IMT-CDMA Multi-Carrier (1X/3X). In the first phase (1X), the cdma2000 standard operates at 144 kbit/s. The second phase (2X) incorporates the 1X standard and adds support to all frequency bands, including 5 and 10 MHz, transmission in circuit and packet mode up to 2 Mbit/s, advanced multimedia and new voice encoders.

1.11 Fourth Generation Systems

The long-term evolution (LTE) technology was proposed as the standard for the fourth generation standardized cellular mobile system. It was developed within the 3 GPP (3rd Generation Partnership Project) as part of the 3 GPP Release eight feature set. Since 2009, LTE mobile communication systems are deployed as an evolution of GSM UMTS and cdma2000 (Roessler et al., 2014).

1.11.1 The Long-term Evolution

The LTE project was initiated in 2004, motivated by the desire to reduce the cost per bit, the addition of lower cost services with better user experience, the flexible use of new and existing frequency bands, a simplified and lower cost network with open interfaces, and a reduction in terminal complexity with an allowance for reasonable power consumption (Erik Dahlman and Beming, 2008; Holma and Toskala, 2009; Khan, 2009).

Other expectations for LTE included a reduction in the packet latency, and an improvement in spectral efficiency, over high-speed packet access (HSPA), of three to four times, in the downlink, and two to three times, in the uplink. Flexible channel bandwidths are specified at 1.4, 3, 5, 10, 15, and 20 MHz in both the uplink and the downlink. This allows LTE to be flexibly deployed in place of other systems, including narrowband systems such as GSM and some systems in the U.S. based on 1.25 MHz (Astély et al., 2009).

Table 1.4 Presents the evolution of UMTS specifications to the LTE standard (Alencar, 2012).

The increase in transmission rate is a notable feature of LTE. Examples of transmission rates for downlink and uplink, for a 20-MHz channel, are shown in Tables 1.5 and 1.6.

Table 1.4   Evolution of UMTS specifications

Version	Date	Main Characteristics
Rel-99	03/2000	UMTS 3.84 M*chip*/s (W-CDMA FDD e TDD)
Rel-4	03/2001	1.28 M*chip*/cps TDD (aka TD-SCDMA)
Rel-5	06/2002	HSDPA
Rel-6	03/2005	HSUPA (E-DCH)
Rel-7	12/2007	High-speed packet access (HSPA)+ (64QAM DL, multiple-input and multiple-output (MIMO), 16QAM UL)
		LTE and SAE feasibility study
		Evolution from EDGE
Rel-8	12/2008	LTE work item – OFDMA air interface
		SAE work item, New IP core network
		3G femtocells, HSDPA dual carrier
Rel-9	12/2009	Multi-standard radio (MSR), HSUPA dual cell
		LTE-Advanced feasibility study, SON
		LTE femtocells
Rel-10	03/2011	LTE-Advanced (4G) work item
		CoMP study, HSDPA with four carriers

Table 1.5 Peak rates for LTE—Downlink, 64-QAM

Antenna	SISO	2×2 MIMO	4×4 MIMO
Peak rate (Mbit/s)	100	172.8	326.4

Table 1.6 Peak rates for LTE—Uplink, single antenna

Modulation	Quaternary Phase Shift Keying (QPSK)	16 QAM	64 QAM
Peak rate (Mbit/s)	50	57.6	86.4

As can be seen, LTE is aimed primarily at low mobility applications in the 0–15 km/h range, in which the highest performance is obtained. The system is capable of working at higher speeds and will be supported with high performance from 15 to 120 km/h and functional support from 120 to 350 km/h. Support for speeds of 350–500 km/h is under consideration within the working groups of the 3rd Generation Partnership Project (3GPP).

2

Cellular Communications Models

The application of communication theory for the wireless environment is a rich subject, that calls for continuing research and involves many diverse aspects of practical and theoretical importance. Mobile and wireless are terms that convey slightly different meanings.

A mobile system is defined as a radio communication network that allows continuous mobility through many cells. Wireless communication, on the other hand, implies radio communication without necessarily requiring the handoff from cell to cell during the conversation (Nanda and Goodman, 1992).

Providing portable radio communication to people on the move has become a major area of research in the past decades (Hashemi, 1991). This should be combined with the increasing interest in avoiding the expensive installation and relocation costs associated with the wired environment (Freeburg, 1991).

Electromagnetic propagation in mobile outdoor and indoor environments is a complex and challenging phenomenon. Multipath dispersion by the various structures of the building results in the reception of a series of echoes for each transmitted pulse. The performance of mobile communication systems for the indoor channel is affected by short-term multipath and large-scale path losses. These problems tend to introduce errors in the transmission and reduce the coverage area (Hashemi, 1991).

Channel characterization is a major research topic in the investigation of such systems, and the usual approaches include modeling the attenuation as a function of frequency, distance, and location, measuring the temporal variation of signal amplitude and phase, use of correlation to assess potential diversity improvements and establishing the impulse response of the channel, in order to measure and characterize the delay spread.

On the other hand, a probabilistic analysis of the wireless indoor communication channel is important to provide the decoder with essential information on the channel state and improve the effectiveness of the communication.

Such modeling must include fading and multipath, as well as interference. Typical methods to combat fading and interference on a channel include the use of burst error correcting codes, equalization, diversity, modulation schemes, filtering, and interleaving. The choice of the best technique depends on the channel characteristics and on the available technology.

2.1 Basic Concepts

When comparing communication channels two fundamental concepts are those of bandwidth and frequency. In practice, the rate at which a source generates information is related to both concepts. The channel bandwidth is directly related to the rate at which information can be sent through the channel, and the channel operating frequency range is related to the carrier modulating frequency (Schwartz et al., 1966).

The choice for the channel that best adapts to these basic concepts is of fundamental importance in the design of a communication system. The choice must be made such that it takes into account the noise introduced by various system components, trying to minimize this noise. It must also be taken into consideration the use of repeaters placed along the signal path in order to amplify the signal level, keeping the signal magnitude at an adequate level so as to compensate for transmission losses (Schwartz, 1970).

The design of a communication system and the choice of channel must take into consideration the fact that the transmission of a large amount of information in a short time interval requires a wide bandwidth system to accommodate the signals. The width of a frequency band appears thus as a fundamental limitation. When signal transmission occurs in real time, the design must allocate an adequate bandwidth for the system. If the system bandwidth is insufficient it may be necessary to reduce the rate of information transmission, thus causing an increase in transmission time.

It must also be taken into account that the equipment design not only considers bandwidth required, but also the ratio (fractional band) between bandwidth and the center frequency in the frequency band. The modulation of a wideband signal by a high-frequency carrier reduces the fractional band and consequently helps to simplify equipment design. In an analog way, for a given fractional band defined by equipment consideration, the system bandwidth can be increased almost indefinitely by raising the carrier frequency.

A microwave system with a 5-GHz carrier can accommodate 10,000 times more information in a given time interval than a system with a 500-kHz carrier frequency. On the other hand, a laser with frequency 5×10^{14} Hz has a theoretical information carrying capacity exceeding that of the microwave system by a factor of 10^5 or, equivalently, can accommodate 10 million TV channels. For this reason communication engineers are continually searching for new sources of high-frequency carriers, as well as for channels that are best matched to such carriers in order to provide wider bandwidths.

As a final comment in this section, for a given communication system and a fixed signal-to-noise ratio (SNR), there is an upper limit defined for the rate of information transmission with a prescribed measure of quality. This upper limit, called the channel capacity, is one of the fundamental concepts of information theory.

Because capacity is a finite quantity, it is possible to say that the design of a communication system is in a certain way a compromise solution involving transmission time, transmitted power, bandwidth, and SNR. All those depend also on the choice of the proper channel, a fact that imposes even more strict constraints due to technological issues (Schwartz, 1970).

2.2 Non-Guided Channels

The main characteristics of non-guided channels are the following.

- Propagation of an electromagnetic wave from the transmitting antenna through free space until it reaches the receiving antenna.
- Irradiated power loss with the square of distance in free space.
- Propagation in the terrestrial atmosphere, causing greater losses due to particles, water vapor and gases, of dimension comparable to the wavelength of the transmitted radiation; attenuation with rain, snow, and other weather factors and attenuation due to obstacles.

The free space non-guided channel can still be subdivided into four categories, according to the wave type, antenna elevation angle, propagation characteristics of the tropospheric and ionospheric layers and the type of link employed, as follows.

- The terrestrial wave can be decomposed into a surface wave and a space wave. The surface wave propagates close to the Earth surface in a manner similar to the wave that propagates in a transmission line. The surface wave follows the Earth surface contours, causing the electric field to bend in respect to the Earth surface, due to signal power losses on the Earth soil due to a relative permittivity and a finite surface conductivity. Mobile communication systems, for example, benefit from the surface wave for signal transmission.

The space terrestrial wave is due to the diffraction of the wave in the propagation media, consisting of a direct wave and a refracted wave.

- For the tropospheric propagation the wave is reflected or refracted in the troposphere, according to the variation of the refraction index of the layers in the latter. It turns out that the signal path is bent due to the refraction index non-homogeneity. When a statistical model is applied to the tropospheric channel the Rayleigh amplitude distribution results.
- The sky wave uses the ionosphere as the natural ionized media for the reflection of radio waves within certain a frequency range and a certain antenna elevation angle.

The ionosphere is the region which extends approximately between 70 and 450 km above sea level and in which the constituent gases are ionized due to the Sun ultraviolet radiation. At such altitudes air pressure is so low that free electrons and ions can exist for relatively long periods of time without recombining to form neutral atoms.

The ionosphere tends to reflect back to Earth waves that originate at ground level and hit the ionosphere, according to some incidence angle.

- In the outer space type of propagation the radiation elevation angle with respect to the earth surface is such that the signal manages to go through the ionospheric layer, penetrating the outer space. This propagation mode is usually employed for satellite transmission.

The noise received by the antenna comes from many distinct sources. The antenna noise is usually characterized by the antenna temperature according to the frequency of operation, since the latter is an important factor to be considered in this type of propagation. It is also possible to obtain curves of minimum and maximum atmospheric noise, also known as static precipitation, originated by the natural occurrence of electrical discharges such as lightening.

Between 50 and 500 MHz, the atmospheric noise becomes negligible, and consequently the cosmic noise, or galactic noise, begins to predominate. This noise exists due to the emission of cosmic sources belonging to our galaxy. Above 1 GHz, the noise contribution due to the phenomenon of atmospheric absorption of atmospheric oxygen becomes important. Heating of the lower atmosphere, where molecular oxygen is found, provides conditions for the occurrence of this type of noise, the absorption peaks of which occur between 22 and 60 GHz.

The combination of the two types of noise mentioned earlier constitutes what is known as sky noise. For the region situated between 1 and 10 GHz, the sky oriented antenna receives a noise power smaller than it would receive if it was working over higher or lower frequencies. This lower power noise level is particularly advantageous in satellite communication systems due to the low-power levels of the processed signals (Gagliardi, 1988).

The adequate use of each channel is a function of the system design for that particular end, as a function of its characteristics; therefore, each communication channel has its importance due to some specific application, each one having advantages and disadvantages. The system design must take into account all those aspects. Usually propagation requirements, bandwidth and cost dictate the solution.

2.3 Effects on the Transmitted Signal

The transmitted signal suffers various effects when traversing the channel. The main effects are the following.

- Filtering – This type of effect tends to reduce the available bandwidth of the modulated carrier, since filtering affects the shape of the modulated waveform causing also phase distortion;
- Doppler – This effect makes the carrier frequency at the reception to differ from that at the transmitter. This effect is due to the frequency deviation provoked by the relative movement between transmitter and receiver. It affects reception of the synchronization signal;
- Fading – This is the name given to the phenomenon causing random variations in the received signal magnitude, in time. This variation has as a reference the value of the received electric field in free space. The causes for fading are found in the propagation media or, in other words, such phenomenon would not be present in the communication links if it were not for the existence between antennas of a media subject to changes in its characteristics;
- Multipath – The received signal is the resultant of the sum of a direct ray between antennas and other rays which follow paths distinct from that of the direct ray. These distinct paths, characterized by multiple paths, originate from refractions and reflections (even if of small intensity), resulting from irregularities in the atmosphere dielectric constant with altitude.

The energy transported by means of such multiple paths is, in general, much less than that associated with the main beam. However, when for some reason (partial obstruction, interference caused by reflection in the terrain, etc.)

the main beam suffers a considerable attenuation, the energy received by means of multiple paths then plays an important role, giving origin to non-negligible interference phenomena.

2.4 The Mobile Communication Channel

Radio propagation in a mobile communication environment, outdoors, or indoors (buildings) is certainly a complex phenomenon. Multipath dispersion caused by the various structures in a building causes a series of echoes for each transmitted pulse.

The performance of mobile communication systems is affected by the effect of multipaths of short duration and by large-scale path losses. These problems tend to introduce errors in transmission and to reduce the area of signal coverage [Hashemi, 1991].

Channel characterization is important in the investigation of the mobile channel. The usual way to address this theme is as follows. Modeling attenuation as a function of frequency, distance, and location; measurement of the time variation of amplitude and phase; use of correlation to assess potential improvements with the the technique of diversity and to establish the channel impulse response, so as to determine the delay spread.

On the other hand, a channel probabilistic analysis is necessary to provide the decoder with essential information about the state of the channel, so as to make communication more effective. Typical methods to combat fading and interference in a channel include the use of burst error correcting codes and interleaving. The choice of the best technique will depend on the channel characteristics.

2.5 Multipath Effects

Fading as caused by multipath, due to reflections on buildings and natural obstacles, for outdoor mobile communication, or on walls, roof and floor, for indoor mobile communication, provokes a series of dips in the received signal spectrum. The pattern thus received represents the signature of selective fading, which can be fairly well predicted by means of an accurate channel analysis.

Fading can also be caused by the movement of people, equipment or cars, such that the resulting interference is the combined effect of multiple access interference (MAI), thermal and impulse noise provoked by electrical equipment. Fading due to transmitter or receiver movement can predominate, in the case of indoor communication, when the indoor objects move relatively slow (Rappaport, 1989).

Many experiments have been performed in the UHF band, between 300 MHz and 3 GHz (Bultitude, 1987; Lafortune and Lecours, 1990). This is due in part to the authorization granted in USA, by the Federal Communications Commission (FCC), for operation with spread spectrum in the 900, 2,400, and 5,725 MHz bands. This is also due to Japan for having chosen the 400 and 2,450 MHz bands for indoor mobile systems (Newman Jr., 1986; Rappaport, 1989).

These frequency bands are usually used in low-capacity voice communication systems, since this application presents less stringent transmission requirements. However, in order to provide enough capacity to cope with the needs of digital transmission systems, further research is needed into the EHF band.

In fact, both the microwave and millimeter wavebands are adequate for data transmission at rates above 10 Mbits/s and coverage of micro or picocells. These frequencies present a series of advantages, usually related to their dual behavior in terms of propagation. They combine properties of UHF frequencies and those of infra-red (IR) light, that is, diffusion, reflection, refraction, occupation of spaces, and are blocked by large objects.

This characteristic allows the signal to fill in the cell space almost completely, with a minimum of signal spillage over neighboring cells, in a manner to provide an adequate company environment, with micro or picocells defined by natural indoor barriers, that is, floor and walls. Mutual interference with existing equipment is normally kept at a minimum level, due to the low-power levels employed or due to the high frequencies in use.

Although interference in indoor mobile communication due to industrial equipment is significant in HF and VHF, it drops quickly above 1 GHz (Rappaport, 1989). Operations at high frequencies allow also the use of much smaller antennas.

Nowadays it is accepted that the indoor mobile channel varies in time and space and is characterized by high-path losses and sudden changes in average signal level. The time variations in the signal statistics are produced mainly by the movement of people and equipment inside a building.

Propagation via multipath is a very often used expression to describe the propagation characteristics in the radio channel. In particular, a signal propagates through different paths in the channel, in which each path presents an associated gain, phase, and delay. Multipath signals recombine in the receiver in a manner that makes the received signal to be a distorted version of that transmitted.

There are three different propagation effects: reflection, diffraction, and scattering. Diffraction occurs when the signal rays bend around obstacles; reflection occurs when the rays collide with hard and even surfaces, and scattering occurs when a ray splits into various rays, after an impact with a hard and uneven surface.

The manner by which multipath signals combine at the receiver leads to the particular type of distortion in the received signal. Two types of recombination may occur, i.e., paths with identical delay will be amplified or attenuated, depending whether the respective signals are in phase or not.

This type of distortion is known as narrowband fading and as a consequence causes bursts of errors. Recombination which do not occur simultaneously or paths with different delays will produce echoes. This type of distortion, called wideband fading, causes pulse spreading which leads to intersymbol interference (ISI).

The effects of fading observed in a mobile channel can be grouped into three different categories: spatial fading, time fading, and frequency selective fading. Spatial fading is characterized by a variation of signal intensity as a function of of distance and occurs in two ways: slow fading and fast fading. Slow fading is characterized as the mean value of the set of signal fluctuations whereas fast fading is characterized by the variation of the signal values around its average value.

The following equation describes slow fading as a function of distance

$$P_{\mathrm{R}} = \alpha P_{\mathrm{T}} e^{-\beta x} \tag{2.1}$$

in which P_{T} is the transmitted signal power, P_{R} is the received signal power, α is the linear attenuation parameter, x is the distance between the transmitter and the receiver, and β is the exponential attenuation factor.

For a perfect outdoor channel (free space) the attenuation factor is $\beta = 2$; however, for an indoor channel (building interior) a strong correlation exists between n and the channel topography. For example, in roomy environments and corridors $\beta < 2$, indicating that signal intensity is greater than that obtained in free space.

This occurs due to a phenomenon called channelling, which produces a kind of waveguide containing the signal. Inside rooms or offices $\beta > 2$ is observed. The signal intensity is smaller than that in free space. This is due to absorption caused by obstacles such as furniture and half walls.

Besides channelling and absorption it was demonstrated that the parameter β depends on three other parameters (Jean-François and Lecours, 1990): the distance between the transmitter and the nearest wall, the distance between the transmitter and any door, and whether the doors are open or closed. Due to the gradual drop in signal level with distance, the characteristics of selective fading are used to calculate coverage areas and to determine the location of sites for placing cells.

Similar to slow fading, fast fading depends also on the terrain topography. Statistical analysis shows that the variations in fast fading are usually of the Rayleigh, Rice, or log-normal type, depending on the terrain topography (Rappaport, 1989). These findings are consistent with what would be expected. That is, if a direct signal is dominant then fading tends to be of the Rice type. On the other hand, if there is no direct signal ray, or whether the signal is dominated by random multipath, then fading will be of the Rayleigh type.

The second type of fading is time fading, which is characterized by the presence of a signal intensity variation, measured at a particular fixed frequency, as a function of time. Other names often employed are narrowband fading, flat fading, and frequency non-selective fading. The time fluctuations in an office environment are in general bursty, while variations in a factory environment are more continuous.

One of the causes for time fading is the physical movement inside the channel, i.e., mobile entities can block signal paths and absorb part of the signal energy temporarily, in a manner so as to create momentary fading. In an outdoor channel rain, water vapor, aerosols, and other substances are energy absorption agents. In the indoor channel people are the primary absorption agents.

The other cause of time fading is the time-varying nature of the channel. Contrasting with the primary source of time fading, physical movement of entities in the channel, the second source can be seen as movement of the channel itself. This movement is caused by changes in propagation characteristics, resulting from changes in room temperature, changes in relative humidity, doors opening and doors closing. Movement inside the channel has a dominant effect on time fading. However, channel movement can affect long-term statistics of the channel fluctuations.

The third type of fading, called frequency selective fading, is characterized by a variation on signal intensity as a function of frequency, also known as wideband fading. Frequency selective fading is the dual of time fading. In time fading the time-varying signal is measured at a fixed frequency while in frequency selective fading the frequency-varying signal is measured in a fixed time instant.

The main cause of frequency selective fading is multipath propagation. At certain frequencies, a combination of multipath signals provokes a reduction in the received signal level, for out of phase signals, and an increase in signal level for in-phase signals. A received signal resulting from a combination of a few components in general will show deep fades, while a signal resulting from a combination of many components in general will show shallow fades.

2.5.1 Statistical Modeling of the Mobile Channel

A complete description of the mobile communication channel would be too complex and beyond hope. Models for the mobile channel must take into account unwanted effects like fading, multipath, as well as interference. Many models have been proposed to represent the behavior of the envelope of the received signal.

A common model to extract statistics of the amplitude of signals in an environment subject to fading is provided by the Rayleigh distribution (Kennedy, 1969). This distribution represents the effect of the amplitude of many signals, reflected, or refracted, reaching a receiver, in a situation in which there is no prevailing component or direction (Lecours et al., 1988). The Rayleigh probability density function is given by Proakis (1990)

$$p_X(x) = \frac{x}{\sigma^2} e^{-\frac{x^2}{2\sigma^2}} u(x), \tag{2.2}$$

with average $E[X] = \sigma\sqrt{\pi/2}$ and variance $V[X] = (2 - \pi/2)\sigma^2$. The associated phase distribution, in this case, is considered uniform in the interval $(0, 2\pi)$. It is possible to approximate with a Rayleigh distribution the amplitude distribution of a set of only six waves with independently distributed phases (Schwartz et al., 1966).

Considering the existence of a strong direct component in the received signal, in addition to multipath components, the Rice distribution is used in this case to describe fast envelope variations in this signal. This line-of-sight component reduces the variance of the signal amplitude distribution, as its intensity grows in relation to the multipath components (Lecours et al., 1988; Rappaport, 1989). The Rice probability density is given by

$$p_X(x) = \frac{x}{\sigma^2} e^{-\frac{x^2+A^2}{2\sigma^2}} I_o\left(\frac{xA}{\sigma^2}\right) u(x), \tag{2.3}$$

in which $I_o(\cdot)$ is the modified Bessel function of order zero and A denotes the signal amplitude. The corresponding variance, for a unit average value, is $V[X] = A^2 + 2\sigma^2 + 1$. The average or expected value is given by

$$E[X] = e^{-\frac{A^2}{4\sigma^2}} \sqrt{\frac{\pi}{2}} \sigma \left[\left(1 + \frac{A^2}{2\sigma^2}\right) I_o\left(\frac{A^2}{4\sigma^2}\right) + \frac{A^2}{2\sigma^2} I_1\left(\frac{A^2}{4\sigma^2}\right) \right], \quad (2.4)$$

in which $I_1(\cdot)$ is the modified Bessel function of order one.

The term $A^2/2\sigma^2$ is a measure of the fading statistics. As the term $A^2/2\sigma^2$ increases, the effect of multiplicative noise or fading becomes less important. The signal pdf becomes more concentrated around the main component. The remaining disturbances show up as phase fluctuations.

On the other hand, signal weakening can cause the main component not to be noticed among the multipath components, originating thus the Rayleigh model. An increment of A, with respect to the standard deviation σ makes the statistics to converge to a Gaussian distribution with average value A (Schwartz, 1970).

Under the conditions giving origin to the Rice distribution, it makes no sense to assume a uniform phase probability distribution. The joint amplitude and phase distribution, which originate the Rice model for amplitude variation, is given by

$$p_{X\Theta}(x, \theta) = \frac{x e^{-A^2/2}}{2\pi\sigma^2} e^{-(x^2 - 2xA\cos\theta)/2\sigma^2} u(x). \quad (2.5)$$

Integrating the above expression for all values of x, the marginal probability distribution θ for the phase results

$$p_\Theta(\theta) = \frac{e^{-s^2}}{2\pi} + \frac{1}{2}\sqrt{\frac{s^2}{\pi}} \cos\theta \, e^{-s^2 \sin^2\theta}[1 + 2(1 - Q(s/\sqrt{2}))\cos\theta], \quad (2.6)$$

in which the function $Q(x)$ is defined in the usual manner

$$Q(x) = \frac{1}{\sqrt{2\pi}} \int_x^\infty e^{\frac{-y^2}{2}} dy. \quad (2.7)$$

The SNR is an auxiliary parameter given by $s^2 = A^2/2\sigma^2$. Expression (2.6) generates a curve with a bell shape for high values of the SNR. For $A = 0$ this probability distribution converges to a uniform distribution (Schwartz, 1970).

Another distribution that has found application for modeling fading in multipath environments is the Nakagami pdf. This distribution can be applied in the situation in which there is a random superposition of random vector components (Shepherd, 1988). The Nakagami probability density function is given by

$$p_X(x) = \frac{2m^m x^{2m-1}}{\Gamma(m)\Omega^m} e^{-\frac{mx^2}{\Omega}} u(x) \tag{2.8}$$

in which $\Omega = P_X$ denotes the received signal average power and $m = \Omega^2/E[(X^2 - E^2[X])^2]$ represents the inverse of the normalized variance of X^2. The parameter m is known as the distribution modeling factor and can not be less than $1/2$. It is easy to show that Nakagami's distribution contains other distributions as particular cases. For example, for $m = 1$ the Rayleigh distribution is obtained.

Finally, the log-normal distribution is used to model certain topographic patterns which appear due to non-homogeneities in the channel, or due to transmission in densely packed spaces or spaces with obstacles (Rappaport, 1989; Hashemi, 1991).

The log-normal distribution is represented by the expression

$$p_R(r) = \frac{1}{\sigma r \sqrt{2\pi}} e^{-\frac{(\log r - m)^2}{2\sigma^2}} u(r) \tag{2.9}$$

has average value $E[X] = e^{\sigma^4/2+m}$ and variance $V[X] = e^{\sigma^4+m}(e^{\sigma^4} - 1)$, and can be obtained directly from the Gaussian distribution by means of an appropriate transformation of variables.

The sequence of arrival times for the various paths $\{\sigma_k\}$ assumes in general a Poisson distribution, given that such sequences are completely random. The Poisson probability distribution function is given by

$$P_k(t) = \frac{(\lambda t)^k}{k!} e^{-\lambda t}, \quad k = 0, 1, 2, \ldots \tag{2.10}$$

in which k represents the number of events counted in the given time interval and λ can be interpreted as the mean rate of occurrence of events (Blake, 1987). The expected value of the Poisson distribution and its variance are both equal to λt.

Exhaustive measurements performed in factory environments led to the conclusion that the time interval between consecutive arrivals $\{t_i = \sigma_i - \sigma_{i-1}\}$ obeys a Weibull distribution with pdf given by (Rappaport, 1989; Hashemi, 1991)

$$p_T(t) = \beta \alpha t^{\beta-1} e^{-\alpha t^\beta} u(t), \tag{2.11}$$

in which α, $\beta > 0$ are parameters of the distribution. For $\beta = 1$ the distribution converges to the Exponential distribution (Blake, 1987).

The Weibull distribution appears when results of radio propagation measurements are plotted on a flat surface with scales adjusted in a manner that the plot of the Rayleigh distribution appears as a straight line with slope equal to -1 (Shepherd, 1988).

Propagation measurements in medium size buildings, however, indicate a tendency of echoes arriving in groups (Saleh and Valenzuela, 1987). This effect is attributed to the building structure, whereby multipath components arriving in groups are due to reflection from objects in the space next to the transmitter or next to the receiver.

The average $E(T)$ and the variance $V(T)$ of the Weibull distribution are given by (Leon-Garcia, 1989)

$$E[T] = \alpha^{-1/\beta}\Gamma\left(1 + \frac{1}{\beta}\right) \tag{2.12}$$

$$V[T] = \alpha^{-2/\beta}\left\{\Gamma\left(1 + \frac{2}{\beta}\right) - \left[\Gamma\left(1 + \frac{1}{\beta}\right)\right]^2\right\}. \tag{2.13}$$

2.5.2 The Two-Ray Model of the Mobile Channel

The two-path model of the mobile channel is based on the diagram shown in Figure 2.1, in which a base station (BS), with an antenna height h_1, transmits to the mobile terminal the antenna height of which is h_2. The distance between the two stations is d (Lee, 1989).

In Figure 2.1, the distances traveled by the direct and by the reflected rays are, respectively

$$d_1 = \sqrt{(h_1 - h_2)^2 + d^2}$$

and

$$d_2 = \sqrt{(h_1 + h_2)^2 + d^2}$$

and it follows that the difference between traveled distances is given

$$\Delta d = d_1 - d_2 = d\left[\sqrt{1 + \left(\frac{h_1 + h_2}{d}\right)^2} - \sqrt{1 + \left(\frac{h_1 - h_2}{d}\right)^2}\right]. \tag{2.14}$$

This formula can be simplified by considering that the mobile unit is far from the BS, i.e., that

$$\frac{h_1 + h_2}{d} \ll 1$$

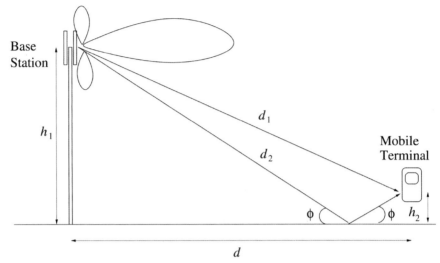

Figure 2.1   A two-path model.

which leads to

$$\Delta d \approx \frac{2h_1 h_2}{d}.$$ (2.15)

The received power is given by

$$P_R = P_0 \left(\frac{\lambda}{4\pi d}\right)^2 |1 + a_\nu e^{j\Delta\phi}|^2$$ (2.16)

in which P_0 represents the transmitted power, $a_\nu = -1$ is the ground reflection coefficient and $\Delta\phi$, the phase difference, is given by

$$\Delta\phi = \beta\Delta d = \frac{2\pi\Delta d}{\lambda} \approx \frac{4\pi h_1 h_2}{\lambda d}.$$

The expression for the received power can then be written as

$$P_R = P_0 \left(\frac{\lambda}{4\pi d}\right)^2 |1 - e^{j\Delta\phi}|^2,$$

which can be simplified even further for $\Delta\phi \ll 1$, resulting in

$$P_R = P_0 \left(\frac{\lambda}{4\pi d}\right)^2 (\Delta\phi)^2 = P_0 \left(\frac{h_1 h_2}{d^2}\right)^2.$$ (2.17)

From Equation (2.17) it can be concluded that the received power, under the model conditions, is proportional to the square of the antenna heights – and decreases with the fourth power of the distance between them.

2.5.3 Two-Ray Model with Frequency Selectivity

Channel characterization is an important research topic in the design of communication systems. In fact, the choice of the best technique for multiple access depends on the channel characteristics.

This section presents a model for the mobile communications environment, which characterizes the channel properties in terms of its frequency response. This model is not intended to be complete but will include also some spectral scattering characteristics which can be useful in the design of communication systems.

A fading model will now be presented, as a function of frequency, which considers the effect of the specular component and the environmental noise. The model is illustrated in Figure 2.2.

It is assumed that the attenuation functions, α and β, have a multiplicative effect on the direct and on the reflect signal, respectively. The net effect of multipath is the introduction of frequency selective fading in the transmitted signal.

The relevance of the analysis of this model comes from the fact that frequency selectivity is the main cause of error bursts in digital communication systems. In this model $s(t)$ represents the transmitted signal, $r(t)$ denotes the received signal, selected among M distinct sequences, and $n(t)$ denotes the environmental noise.

All stochastic processes considered in this model are assumed to be stationary, at least wide sense stationary. As usual, the noise is assumed to be independent of the transmitted signal, and having two components. The first component $n_A(t)$ takes into account equipment noise and environmental noise, considered as an additive white Gaussian noise (AWGN) process with zero mean. The second component $n_I(t)$ represents the interference due to the remaining $M - 1$ signals.

$$n(t) = n_A(t) + n_I(t). \tag{2.18}$$

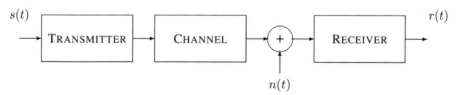

Figure 2.2   Channel model including frequency selectivity.

Considering the assumed independence between the noise processes and the transmitted signals, in the worst case, it follows that

$$S_N(w) = S_A(w) + S_I(w) \tag{2.19}$$

or

$$S_N(w) = S_o + (M - 1)S_S(w) \tag{2.20}$$

in which S_o represents the noise power spectral density – PSD, and $S_S(w)$ is the transmitted signal PSD. For simplicity it is assumed that the various transmitted signals are statistically similar.

The time delay σ for the reflected signal is assumed to be constant, even if that may appear to be a rather restrictive assumption for a mobile system. This consideration is based on the fact that the distance traveled by the user is usually much less than the average distance between the user and the BS-transmitting antenna.

The consequence of a time-varying delay is a change in the separation between fading valleys in the spectrum of the received signal. This effect can be significant for short distances. The received signal is given by

$$r(t) = \alpha(t)s(t) + \beta(t)s(t - \sigma) + n(t). \tag{2.21}$$

Three important cases are considered in this section. In the first case the attenuation functions are considered time independent. In the second case, these functions are considered to be time-varying but deterministic.

Finally, in the third case it is considered that the two attenuation functions are of a random nature. The autocorrelation function and the power spectral density of the received signal are computed in order to provide information about the channel behavior. The autocorrelation and the power spectral density for the preceding equation are given, respectively, by

$$R_R(\tau) = E[r(t)r(t + \tau)] \tag{2.22}$$

and

$$S_R(w) = \int_{-\infty}^{\infty} R_R(\tau)e^{jw\tau}dw. \tag{2.23}$$

Consider constant attenuation functions $\alpha(t) = \alpha$ and $\beta(t) = \beta$ for the first case. By substituting these values in the previous two equations it follows that

$$S_R(w) = \int_{-\infty}^{\infty} R_R(\tau)e^{jw\tau}dw. \tag{2.24}$$

and

$$S_R(w) = [\alpha^2 + \beta^2 + 2\alpha\beta \cos w\sigma] S_S(w) + S_N(w). \qquad (2.25)$$

The square of the channel transfer function modulus can be obtained directly from the preceding equation, that is,

$$|H(w)|^2 = [\alpha^2 + \beta^2 + 2\alpha\beta \cos w\sigma]. \qquad (2.26)$$

From Equation (2.26) it is easy to grasp the role of the parameters α, β, and σ with respect to channel behavior as a function of attenuation and frequency selectivity. For example, σ controls the separation between valleys in the channel transfer function, while α and β control the depth of these valleys. Figure 2.3 illustrates the calculated transfer function.

As a second example, consider that the attenuation functions are time-varying, however of a deterministic nature. Again, introducing the respective parameters in Equations (2.22) and (2.24) leads to

$$R_R(\tau) = \alpha(t)\alpha(t+\tau)R_S(\tau) + \beta(t)\beta(t+\tau)R_S(\tau) + R_N(\tau)$$
$$+ \alpha(t)\beta(t+\tau)R_S(\tau-\sigma) + \alpha(t+\tau)\beta(t)R_S(\tau+\sigma) \qquad (2.27)$$

and

$$S_R(w) = [\alpha(t)A(w)e^{jwt}$$
$$+ \beta(t)B(w)e^{jwt}] \times S_S(w) + [\alpha(t)B(w)e^{jwt}] \times [S_S(w)e^{-jw\sigma}]$$
$$+ [\beta(t)A(w)e^{jwt}] \times [S_S(w)e^{jw\sigma}] + S_N(w). \qquad (2.28)$$

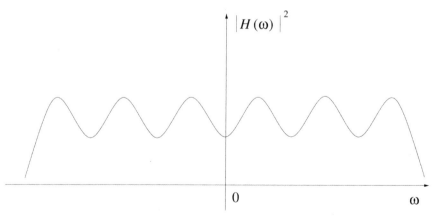

Figure 2.3 Channel signature, or transfer function, for a channel with multipath.

Finally, assuming random parameters $\alpha(t)$ and $\beta(t)$, leads to

$$R_R(\tau) = [R_A(\tau) + R_B(\tau)]R_S(\tau) + R_{AB}(\tau)R_S(\tau - \sigma)$$
$$+ R_{BA}(\tau)R_S(\tau + \sigma) + R_N(\tau) \tag{2.29}$$

and

$$S_R(w) = [S_A(w) + S_B(w)]^* S_S(w) + S_{AB}(w) \times [S_S(w)e^{-jw\sigma}]$$
$$+ S_{BA}(w)^*[S_S(w)e^{jw\sigma}] + S_N(W) \tag{2.30}$$

in which $R_A(\tau)$ and $R_B(\tau)$ represent, respectively, the autocorrelation of $\alpha(t)$ and $\beta(t)$. An analogous notation follows for the spectrum.

2.5.4 Effect of Multiple Rays

A more accurate model can be devised, by considering a large number of rays for the transmitted signal. The general expression for the signal received by one of the users is given by

$$r(t) = \sum_{k=1}^{K} \alpha_k s(t - \sigma_k). \tag{2.31}$$

The attenuation parameters α_k are considered dependent only on the path followed by the signal, l_k, besides dependency on the time instant. The delays, σ_k, are also functions of l_k.

It is important to note that a distinction must be made between models that describe the signal intensity as a function of time and those that describe the signal intensity as a function of the distance traveled. The first is useful for computing the error probability and the margin of operation, while the second is appropriate to determine the coverage area and co-channel interference.

A simplification results when independence between the effects of distance and time is assumed. That implies writing the probability density function for the attenuation parameters as $p(\alpha(t, l_k)) = p(\alpha(t))p(\alpha(l_k))$.

The channel transfer can be obtained directly from the previous equation as

$$h(t) = \sum_{k=1}^{K} \alpha_k \delta(t - \sigma_k). \tag{2.32}$$

The Fourier transform of the previous expression leads to the channel transfer function

$$H(w) = \sum_{k=1}^{K} A_k e^{jw\sigma_k} \tag{2.33}$$

in which A_k represents the Fourier transform of α_k.

In order to consider the effects of phase displacement of the reflected signals, it is more convenient to use the complex notation to represent the received signals. This leads to the low-pass model for the transfer function

$$h(t) = \sum_{k=1}^{K} \alpha_k \delta(t - \sigma_k) e^{j\theta_k} \tag{2.34}$$

in which $\{\theta_k = w_c \sigma_k\}$ represent the phase rotations and w_c denotes the carrier frequency (Schwartz et al., 1966). All parameters involved are considered to be random variables obeying their respective distributions, discussed earlier.

Obtaining the $\{\alpha_k\}$ parameters is of great importance when modeling the mobile channel. However, two of the derived parameters are useful when describing global characteristics of the propagation profile in multipath (Saleh and Valenzuela, 1987; Pahlavan et al., 1989). These are the multipath power gain

$$G_{\mathrm{M}} = \sum_{k=1}^{K} \alpha_k^2. \tag{2.35}$$

and the RMS delay spread

$$\sigma_{\mathrm{RMS}} = \sqrt{E[\sigma^2] - E^2[\sigma]}, \tag{2.36}$$

in which

$$E[\sigma^n] = \frac{\displaystyle\sum_{k=1}^{k} \alpha_k^n \sigma_k^n}{\displaystyle\sum_{k=1}^{k} \alpha_2^k}, \quad n = 1, 2 \tag{2.37}$$

or

$$E[\sigma^n] = \frac{1}{G_{\mathrm{M}}} \sum_{k=1}^{K} \alpha_k^n \sigma_k^n. \tag{2.38}$$

The power gain is useful to estimate the system SNR. The RMS delay spread is a good measure of multipath scattering, besides being related to the performance degradation caused by ISI. As a consequence, the RMS delay spread limits the maximum signaling rate allowed for a given error rate.

Values found in the literature situate the RMS delay spread in the 20–50 ns range, for small office buildings of small to medium size (Saleh and Valenzuela, 1987), between 30 and 300 ns for factory environments (Pahlavan et al., 1989; Rappaport, 1989), and up to 250 ns for large office buildings (Devasirvatham, 1984).

Based on the previous values, it is estimated that digital transmission rates above 400 kbit/s may not be feasible for an error probability of 10^{-3} or smaller in the above referred building (Devasirvatham, 1984). It should be observed that measurements were performed in the 800 MHz to 1.5 GHz frequency band.

2.5.5 Time-Varying Channels

An appropriate channel model must take into account three fading effects: spatial, temporal, and frequency selective. The spatial fading strongly depends on the topography, indicating that it is unique in each case, that is, a model designed for a building may not be adequate for use in another building. Therefore, the dependency with distance will not be included in the channel models to be considered in the sequel. The variables considered will be time and frequency (Thom, 1991).

The frequency response of a time-varying channel is denoted by $h(\tau, t)$. The time-varying frequency response is obtained by means of the Fourier transform of the impulse response with respect to the variable τ

$$H(\omega, t) = \int_{-\infty}^{\infty} h(\tau, t) e^{-j\omega\tau} \, d\tau. \qquad (2.39)$$

Both $h(\tau, t)$ and $H(\omega, t)$ are random processes due to variations in t, representing time. From the variables τ and t other functions can be derived, by means of the autocorrelation function or the Fourier transform of $h(\tau, t)$, i.e., $H(\omega, t)$.

The autocorrelation function of the impulse response $h(\tau, t)$ is defined as

$$R_H(\tau_1, \tau_2; t_1, t_2) = E[h^*(\tau_1, t_1) h(\tau_2, t_2)]. \qquad (2.40)$$

Assuming that the random variables are wide sense stationary, which is valid for symmetric time variations, it follows that the autocorrelation will depend on the difference $\sigma = t_2 - t_1$ only. This model is known as the Wide Sense Stationary (WSS) model. Equation (2.40) then becomes

$$R_h(\tau_1, \tau_2; \sigma) = E[h^*(\tau_1, t_1)h(\tau_2, t_1 + \sigma)]. \tag{2.41}$$

The autocorrelation of the frequency response $H(\omega, t)$ is given by

$$R_H(\omega_1, \omega_2; t_1, t_2) = E[H^*(\omega_1, t_1)H(\omega_2, t_2)]. \tag{2.42}$$

The dual of the wide sense stationary model is the Uncorrelated Scattering (US) model (Bello, 1963). For the Wide Sense Stationary (WSS) model the autocorrelation depends on the time difference $\sigma = t_2 - t_1$ only. A situation analogous to this occurs when the autocorrelation in Equation (2.42) for the US model depends on the frequency difference $\nu = f_2 - f_1$ only. Thus, Equation (2.42) becomes

$$R_H(\nu; t_1, t_2) = E[H^*(\omega_1, t_1)H(\omega_1 + \nu, t_2)]. \tag{2.43}$$

A third possible model, which is the simplest of them all, is that model which is Wide Sense Stationary Uncorrelated Scattering (WSSUS). This type of channel model has the properties of channels already described. In other words, the autocorrelation in Equations (2.40) and (2.42) depend both of σ and ν. Equation (2.42) then becomes

$$R_H(\nu; \sigma) = E[H^*(\omega_1, t_1)H(\omega_1 + \nu, t_1 + \sigma)]. \tag{2.44}$$

In this case, $R_H(\nu; \sigma)$ is called autocorrelation function spaced in time and frequency. By making $\sigma = 0$, the following function results

$$R_H(\nu; \sigma)|_{\sigma=0} = R_H(\nu), \tag{2.45}$$

which is the frequency correlation function. $R_H(\nu)$ measures the degree of correlation between different frequencies. A measure of uncorrelatedness is obtained when $\nu = \omega_c$ such that $R_H(\omega_c) = 0$. Frequencies separated by ω_c will be uncorrelated. The resulting parameter ω_c is called correlation bandwidth.

Thus, if a transmitted signal bandwidth W is narrower than the correlation bandwidth, i.e., if $W < \omega_c$, then that signal frequency content is highly correlated and all its spectrum will suffer similar phase and amplitude variations. This condition is referred to as non-selective fading.

On the other hand, if a transmitted signal bandwidth W is wider than the correlation bandwidth, i.e., if $W > \omega_c$, then the time variation for some frequencies will be independent of the time variation of other frequencies, and that characterizes what is called frequency selective fading.

Other functions can be derived from the relationship between $h(\tau, t)$ and $H(\omega, t)$. Since $h(\tau, t)$ and $H(\omega, t)$ are a Fourier transform pair, it follows that $R_h(\tau_1, \tau_2; t_1, t_2)$ and $R_H(\omega_1, \omega_2; t_1, t_2)$ are also a Fourier transform pair, i.e.,

$$R_H(\omega_1, \omega_2; t_1, t_2) = \int\limits_{-\infty}^{\infty} \int\limits_{-\infty}^{\infty} R_h(\tau_1, \tau_2; t_1, t_2) e^{j(\omega_1 \tau_1 - \omega_2 \tau_2)} \, d\tau_1 \, d\tau_2.$$

(2.46)

Thus, for a WSSUS channel, the replacement of $\sigma = t_2 - t_1$ and $\nu = w_2 - w_1$ in Equation (2.46), after some manipulation leads to

$$R_H(\tau_1, \tau_2; \sigma) = \frac{1}{2\pi} \int\limits_{-\infty}^{\infty} R_H(\nu; \sigma) e^{j\tau_1 \nu} \, d\nu \, \delta(\tau_2 - \tau_1)$$

$$= R(\tau_1, \sigma) \delta(\tau_2 - \tau_1),$$

(2.47)

in which

$$R(\tau_1, \sigma) = \frac{1}{2\pi} \int\limits_{-\infty}^{\infty} R_H(\nu; \sigma) e^{j\tau_1 \nu} \, d\nu$$

(2.48)

$R(\tau_1, \sigma)$ is the Fourier transform of the autocorrelation of the channel frequency response. This function is called the *delay cross-power spectral density function*. For $\sigma = 0$ it follows that

$$R(\tau, \sigma) \big|_{\sigma=0} = R(\tau)$$

(2.49)

and Equation (2.47) becomes

$$R_H(\tau_1, \tau_2) = R(\tau_1) \delta(\tau_2 - \tau_1).$$

(2.50)

The power impulse response $R(\tau)$, in Equation (2.49), expresses a measure of the channel wideband fading characteristics. From Equations (2.45), (2.48) and (2.46), it can be seen that $R(\tau)$ is the inverse Fourier transform of the frequency correlation function $R_H(\nu)$, that is,

$$R(\tau) = \frac{1}{2\pi} \int\limits_{-\infty}^{\infty} R_H(\nu) e^{j\tau\nu} \, d\nu.$$

(2.51)

An important parameter which can be extracted from $R(\tau)$ is the delay spread σ_{RMS}. This result is obtained by making $\tau = \sigma_{RMS}$, in a manner that $R(\sigma_{RMS}) = 0$. Considering that $R(\tau)$ and $R_H(\nu)$ are Fourier transform pairs, then the delay spread and the correlation bandwidth are related as

$$\sigma_{RMS} = \frac{1}{\omega_c}. \tag{2.52}$$

Two other functions can still be obtained from the WSSUS channel. Differing from the previous functions, which focus on the channel frequency variations, the following functions focus on the channel time variations. The first function is obtained by making $\nu = 0$ in the correlation function spaced in frequency and time $R_H(\nu; \sigma)$. The result is

$$R_H(\nu; \sigma)|_{\nu=0} = R_H(\sigma). \tag{2.53}$$

Thus, $R_H(\sigma)$ is called the *correlation function spaced in time* and describes how fast the channel fades. A useful parameter which can be derived from $R_H(\sigma)$ is the coherence time t_c, which results by making $\sigma = t_c$, such that $R_H(t_c) = 0$.

High values of t_c indicate a channel which is slow in terms of time variations, while small values of t_c indicate a channel with fast variations with respect to time.

The second function is obtained by means of the Fourier transform of $R_H(\sigma)$ with respect to σ, i.e.,

$$S(\lambda) = \int_{-\infty}^{\infty} R_H(\sigma)e^{-j\lambda\sigma} \, d\sigma. \tag{2.54}$$

This function is called the Doppler power spectrum. A useful parameter derived from $S(\lambda)$ is the Doppler bandwidth W_D, which measures the amount of Doppler scattering or movement within the channel. Since the echo power spectrum and the correlation function spaced in time constitute a Fourier transform, then the Doppler bandwidth and the coherence time are related as

$$W_D = \frac{1}{t_c}. \tag{2.55}$$

Consequently, a large Doppler bandwidth indicates a channel with fast time variations, while a narrow Doppler bandwidth indicates a channel with slow time variations.

One last function will now be derived from the WSSUS channel. This function results by considering the two-dimensional Fourier transform of the correlation function spaced in time and frequency $R_H(\nu; \sigma)$, i.e.,

$$S(\tau; \lambda) = \int\limits_{-\infty}^{\infty} \int\limits_{-\infty}^{\infty} R_H(\nu; \sigma) e^{-j\lambda\sigma} e^{j\tau\nu} \, d\sigma \, d\nu. \qquad (2.56)$$

In this case $S(\tau; \lambda)$ is called the channel scattering function. If the channel transfer function is not known in advance, some form of combating its effects must be found. Equalization is the usual procedure for obtaining a flat amplitude response and a linear phase response, which characterize the ideal channel.

An adaptive equalizer is designed from observations of the channel output, for a given input known as training sequence. Alternately, the output sequence can be the channel response to a carrier modulated by a random data sequence. The former approach is used in point-to-point communication systems while the latter is more adequate for broadcast (Blahut, 1990).

One of the first studies about the convergence of equalizers showed that the use of isolated test pulses as the training signal is sub-optimum when compared with the use of pseudo-random sequences, or the data signal itself (Gersho, 1969) as training signals. More complete essays on equalization can be found in (Macchi et al., 1975) and (Qureshi, 1985).

As discussed in this chapter, propagation of signals in mobile channels is subject to severe frequency fading. In an indoor channel presenting a high delay spread, performance will seriously suffer with ISI. Adaptive equalization is efficient against this type of degradation. The final result, as a consequence of improvement in performance, is an increase in the channel signaling rate (Valenzuela, 1989).

A few improvements can be done to the previous model, in order to make it suitable for indoor use. The proposed model is based on the assumption that the indoor channel can be modeled as a WSSUS channel (Yegani and Mcgillen, 1991). This model concentrates on the channel impulse response $h(\tau, t)$, in which the parameters of the model are related to the propagation parameters.

It follows that all channel functions derived from Equations (2.44), (2.45), (2.51), (2.53), (2.54) and (2.56) can be derived from the knowledge of $h(\tau, t)$.

The time varying channel impulse response is given by

$$h(\tau, t) = \sum_{k=0}^{K} \alpha_k(t)\delta(\tau - \sigma_k(t))e^{-j\theta_k(t)}, \qquad (2.57)$$

in which t denotes the sampling instant and $\alpha_k(t)$, $\theta_k(t)$ and $\sigma_k(t)$ are random variables representing the time varying attenuation, phase and delay in the k-th channel path, respectively.

The statistical distributions of multipath parameters are the following.

- The variable $\alpha_k(t)$ has a Rayleigh, Rice, or log-normal distribution, depending on the terrain topography. For indoor channels usually the Rice distribution is considered while the Rayleigh distribution is frequently used for outdoor channels.
- The variable $\theta_k(t)$ is uniformly distributed between 0 and 2π radians if $\alpha_k(t)$ is Rayleigh or Rice distributed.

Most of the recent research has concentrated on the $\alpha_k(t)$ distribution and a little on the $\theta_k(t)$ distribution, in which a uniform distribution is usually assumed. Far less attention has been given to the $\sigma_k(t)$ distribution.

2.6 Urban Area Propagation Models

An estimate of signal attenuation, as seen in previous sections, is done either by stochastic analysis or by an empirical formulation based on experimental data. Okumura's model and the multiple ray models are employed for the analysis of propagation in urban areas (Yacoub, 1993a).

2.6.1 Propagation in Mobile Systems

An empirical model for propagation prediction for mobile communications was published by Okumura, who performed various field intensity measurements in the urban area of Tokyo and also in a suburban geographic area (Lee, 1989).

The measurements were performed in the 150 MHz to 20 GHz frequency range, with an effective transmitting antenna height varying between 30 and 1000 m, over distances varying from 1 to 100 km. The height of the mobile unit antenna varied from 1 to 10 m.

The following equations, that are listed for reference only, compare various signal propagation models, for the case in which frequency selectivity is not considered, that is, flat fading is assumed. The parameters P_T e P_R represent, respectively, the transmitted and the received power, d is the distance between antennas, G_T and G_R are the respective gains for the transmitting antenna with height h_B, and receiving antenna with height h_M.

- Free space propagation

$$\frac{P_R}{P_T} = \left(\frac{\lambda}{4\pi d}\right)^2 (G_T G_R) \qquad (2.58)$$

- Propagation in a path that includes ground reflection

$$\frac{P_R}{P_T} = \left(\frac{h_B h_M}{d^2}\right)^2 (G_T G_R) \qquad (2.59)$$

- Propagation in an urban environment

$$P_R = P_T - A_M(f, d) + H_B(h_B, d) + H_M(h_M, f) \qquad (2.60)$$

- Propagation a suburban area

$$P_R = P_T + K_{SO} + K_{SP} + K_{Terr} \qquad (2.61)$$

The coefficients $A_M(f, d)$, $H_B(h_B, d)$ e $H_M(h_M, f)$ represent the effect of attenuation or gain in an urban environment, as a function of distance, frequency of operation, and antenna height. The parameters K_{SO}, K_{SP} e K_{Terr} represent the effect of the type of terrain in a clear area.

2.7 Propagation Loss

The irregular nature of the terrain, the types of architectonic structures, the atmospheric changes and the foliage change of the trees difficult the prediction of propagation losses, for radio communication systems (Yacoub, 1993a).

Therefore, it is important to study and plan the cellular system taking into account all the information that can be obtained from the environment, in order to develop mathematical and computational models for cell coverage from the radiobase stations.

Several general results, obtained from prediction models, differ because of the artificial nature of man-made structures, for urban and suburban geography, the nature of open areas, such as mountains, lakes, forests, and plain terrain (Macario, 1991; Stuber, 1991).

The planning of a cellular system is based on the signal coverage, evaluated by means of prediction models, telephone traffic, estimated from queuing theory, automobile traffic, and people circulation. The task of obtaining full coverage in a certain region is a difficult one, and the regulatory agencies usually stipulate that the operators must cover 90% of the area, 90% of the time.

Many traditional models, commonly used to predict coverage, are empiric and rely on path propagation curves that estimate the attenuation of the radio signal between the radiobase station and the mobile device. Some empirical models, found in the literature, do not take into account the terrain characteristics, which produces discrepancies between the measured and predicted values (Lee, 1990).

The observation of 68% of the obtained values, for a standard deviation ratio of 6 to 8 dB, indicate a good match between the predicted and measured results. But it is not advisable to rely on such models for cellular communication systems design, due to the high uncertainty degree of the prediction (Lee, 1993).

In the empirical model, the prediction curve inclination and the intersection point are obtained from measurements taken at a typical propagation environment. Several computational tools are available to the operators, that evolved from the original models of Hata and Lee, to plan the cellular coverage areas, the co-channel interference, and the channel capacity.

The analytical models, on the other hand, explain the path losses in certain environments as the result of signal reduction due to free space propagation loss plus scattering, multiple diffraction in surfaces, and shadowing effects.

For such models, the integration of multiple dimensions to compute the signal attenuation, as a result of multiple reflections, diffractions on different surfaces, is difficult to implement. But, simpler models can be used to obtain quick estimates of a certain coverage area.

2.7.1 Near Field Propagation Model

In cellular planning, it is important to guarantee the system capacity in order to ensure the minimum power necessary to provide acceptable levels of the received signal at the mobile terminals. The transmission effects in adjacent areas must be estimated, to identify the limiting zones, where the signal level is minimum and the ones where there is a possibility of signal interference.

Therefore, it is important to choose an adequate prediction model for the cell planning. This section presents the description of a new propagation model for cell planning.

Curves of signal power strength are forecasted by prediction models and used in the cellular planning. However, for points near the BSs the prediction models usually leads to power levels higher than real measurements.

This problem has been detected in theoretical propagation model as well as the empirical ones, which are based on campsite measurements. It is worth to mention that these models assume the use of omnidirectional antennas.

This approach is useful because it allows drive-test in the whole coverage area. However, this sort of antenna leads to a high signal strength near the BSs because the power propagates in all directions with the same strength (de Oliveira et al., 2006).

Inside an area delimited by a radius of 1 km, approximately, the omnidirectional antennas are limited to transmit in the horizontal plane. Therefore, the signal received by the mobile unit is additionally reduced as a function of the antenna elevation angle, as can be observed in Figure 2.4.

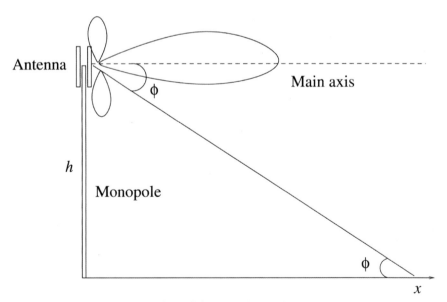

Figure 2.4   Elevation angle.

For the near field propagation, the statistical influence of the streets is reduced, but it is important for medium to far field propagation, in which case it can introduce an additional loss between a0 and 20 dB on the received signal (de Oliveira, 2004).

2.7.2 Equation for the Near Field

By using campsite measurements and heuristic methods, the attenuation model must satisfy some criteria:

- A small increase for small distances and a rapid decrease for long distances;
- The complexity must be low to permit a computational implementation;
- The equation must be in product form in order to fit the decibel format.

The first criterion is established by observing the curves in Figure 2.5, which gives the campsite measurements obtained from a three-sector cell in a mobile system. The second criterion is based on the necessity of implementation of the model in a prediction coverage software.

The software determines de power of the highest number of points of the region which is intended to be a mobile service coverage area. The last criterion comes from the fact the decibel is largely used in propagation engineering.

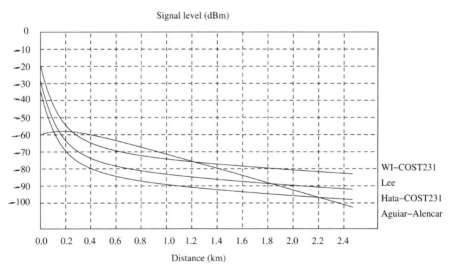

Figure 2.5 Comparison between the Hata–COST231, COST231 Walfish–Ikegami (COST231-WI) models, and the model with the correction coefficient.

This way, the equation which describes the product-behavior of the new model is converted into a set of additions. The following equation satisfies the previous criteria

$$P_R = P_T \left(\frac{x}{x_0} \right) 10^{-\beta \frac{x}{x_0}}. \tag{2.62}$$

The previous expression can be converted to decibel units,

$$P_R = P_T + 10 \log \left(\frac{x}{x_0} \right) - 10\beta \frac{x}{x_0}. \tag{2.63}$$

in which the power P_R is received at x, which is the distance between the transmitter and the receiver. The exponent β corresponds to the path loss. The parameter x_0 represents the initial distance, near the site, and P_T is the transmitted power measured at distance x_0.

The cell site proximity can offset the reception power level, as the mobile unit moves in a radius of 1 km from the BS. As the user moves away from the station, the proximity effects become negligible, and the long term attenuation dominates.

For the measurement of an electromagnetic field at the ground level, it is necessary to adjust the obtained values using a correction coefficient. The models found in the literature usually model the field behavior at the transmitting antenna level (Aguiar et al., 20).

For a station monopole of height h, consider a line that connects the antenna to a measurement point at the ground level, at a distance x from the monopole. The monopole base, the antenna and the measurement point are the vertices of a rectangle triangle, as shown in Figure 2.4.

Assume that the angle at the antenna vertex of the triangle is θ. Therefore, the complement of the angle ϕ is the one between the main axis and the line that connects the antenna and the point x, at the ground level, that is

$$\phi = \frac{\pi}{2} - \theta.$$

On the other hand, the tangent of θ is

$$\tan \theta = \frac{x}{h},$$

in other words,

$$\theta = \arctan \frac{x}{h}.$$

Considering a short dipole, for example, for which the dimension is smaller than the wavelength of the transmitted carrier, ($l \ll \lambda$), the equation for the antenna radiation pattern is

$$g(\phi) = g\cos^2(\phi), \tag{2.64}$$

in which g is the antenna gain ($g \approx 1.5$), and the beamwidth is 90^0.

Therefore, the formula for the field intensity at the ground level is

$$g(\theta) = g\cos^2(\frac{\pi}{2} - \theta),$$

or,

$$g(x) = g\cos^2\left[\frac{\pi}{2} - \arctan\frac{x}{h}\right]. \tag{2.65}$$

The correction coefficient must be incorporated in the attenuation model, to compute the electromagnetic field at a certain distance from the antenna. For the far field, the argument into brackets tends to zero, and $(x \to \infty)$, $g(x) \to g$, as expected, because the measurement point is located at the maximum of the antenna main lobe (Aguiar et al., 2004).

Figure 2.5 illustrates a comparison between the Hata–COST231, and COST231 Walfish–Ikegami (COST231-WI) models, the reference curve (Lee model) and the model with the correction coefficient (Aguiar–Alencar model) (de Oliveira, 2004). The correction coefficient is necessary for a distance smaller than 500 m, which is equivalent to a typical cell radius.

In order to evaluate the performance of the Aguiar–Alencar model, some measurements of the signal strength were performed for a three-sector site, located in the urban area of Manaus, State of Amazonas, Brazil. From the collected data, it was possible to determine the statistical parameters and the corresponding standard deviation between the measured data and the losses predicted by the new model.

The Aguiar–Alencar model produced a standard deviation $\sigma = 9.80\,\mathrm{dB}$, compared to $\sigma = 9.89\,\mathrm{dB}$, and $\sigma = 9.98\,\mathrm{dB}$, respectively, for the the Hata–COST231, and COST231-WI models. It is important to mention that the standard deviation for the three models are close to the results found in the literature, considering urban areas (Andersen et al., 1995).

The near field modelling requires a more complex analysis, because of the oscillating nature of the electromagnetic field for short distances, as can be observed in Figure 2.6, as proposed in (Gomes, 2001).

2.7.2.1 Propagation plots

Using a suburban area, as an example, the signal level measured at a distance of 1 km from the BS is -61.7 dBm, which is due to a set of parameters, including the antenna height of 30 m. If the antenna height is increase to 60 m, a gain of 6 dB is obtained. Between 60 and 120 m, an additional gain

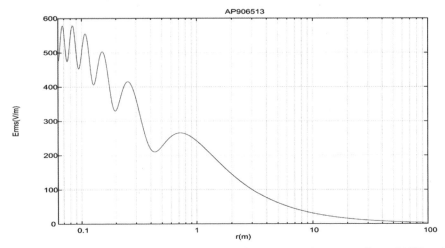

Figure 2.6   Electromagnetic field in the vicinity of the antenna, for a power $P_T = 20$ W, and antenna gain $g = 2$ dBi. The carrier frequency is $f = 900$ MHz.

Table 2.1   Attenuation data for different heights of a specific antenna

Antenna Height, h_1 (m)	Incidence Angle, θ (grades)	Elevation Angle, ϕ (grade)	Attenuation, α (dB)
90	30.4	29.6	21
60	21.61	20.75	16
30	11.77	10.72	6

of 6 dB is achieved. From a height of 120 m and up, the received signal at the mobile terminal is equivalent to that obtained for free space.

The antenna pattern is isotropic in the horizontal plane, also called azimuthal plane. The signal improvement with the antenna height can be seen in Table 2.1. For a distance $d = 100$ m, and for a mobile terminal positioned at 3 m above the ground, the incidence and the elevation angles are 11.77° e 10.72°, respectively.

2.7.3 Far Field Propagation Model

The advantage of large cell sites is the coverage of large areas, mainly for noise limited systems, in which different frequencies are used to cover distinct regions. But, it is necessary to be aware of the propagation phenomena that occur at large distances.

A noise limited system can become an interference limited system as the traffic increases. The interference is not only due to the existence of adjacent channels, but also a result of the effect of propagation at a certain distance.

A circular area, with a radius of 80 km, is affected by a low atmosphere phenomenon that causes a terrestrial wave. This effect is stronger over sea water, because of the particular atmospheric conditions over the ocean, which depends on the altitude. This results in a fading effect, as the wave goes up or down, as it propagates.

Tropospheric waves are important in the 800 MHz range, and the signal can propagate up to 320 km from the BS, due to abrupt changes in the effective dielectric constant over the troposphere. This constant changes with the temperature, and decreases at a rate of $6.5°$ to $7.0°$ C/km with the altitude.

The troposphere contains several gases, such as oxygen, nitrogen, and carbon dioxide, aside from water vapor, aerosols, and, eventually, rain and snow. Therefore, the physical behavior of this atmospheric layer is usually described by three parameters: atmospheric pression, temperature, and water vapor pression.

Regarding the radio waves, the main phenomena that occur during propagation through the atmosphere are:

- Wave refraction.
- Energy absorption by the oxygen and water vapor.
- Energy absorption by the aerosols.
- Influence of precipitations.

2.7.4 Path Propagation Loss

Path propagation loss is an important metric to assess the quality of radio propagation in a wireless channel, defined as

$$l \triangleq \frac{P_R}{P_T},$$

in which (P_T) is the transmitted and (P_R) is the received power.

The propagation loss, in dB, is

$$L = -10 \log l = -10 \log P_R + 10 \log P_T. \tag{2.66}$$

2.7.5 Free Space Propagation Loss

The ration between the received and the transmitted power, for free space propagation conditions, is given by

$$\frac{P_R}{P_T} = G_T G_R \left(\frac{\lambda}{4\pi d}\right)^2, \tag{2.67}$$

in which λ is the wavelength, and G_T and G_R are the respective antenna gains at the transmitting and receiving sides.

The propagation loss, in dB, is written as

$$L = -10\log G_T - 10\log G_R - 20\log \lambda + 20\log d + 21,98. \tag{2.68}$$

For isotropic antennas $(G_T = G_R = 1)$, and the frequency given in MHz and the distance given in kilometers, one obtains

$$L = 20\log f + 20\log d + 32.44\,(\text{dB}). \tag{2.69}$$

2.7.6 Propagation Loss Along the Terrestrial Surface

Consider the electromagnetic propagation in a plane surface. The transmitted signal has several paths to reach the receiving antenna:

- Direct path.
- Indirect path, cause by a reflection of the wave on the ground.
- Indirect path, consisting in a surface wave.
- By means of secondary paths.

The received signal is a combination of all the multipaths, and the resulting power is the sum of the individual ones. The direct path signal power is obtained using the formula for free space propagation loss. The reflected wave power is given by the free space propagation equation subject to an additional attenuation, which is equivalent to the ground reflection coefficient ρ.

In addition, the reflected signal has a phase displacement $\Delta\varphi$ because of the indirect path. The terrestrial wave is produced by the signal that is absorbed by the soil. The fraction of the absorbed signal is given by $(1 - \rho)$, which corresponds to the reflected signal multiplied by an attenuation factor A.

Therefore, the ratio between the powers of the received and transmitted signals is

$$\frac{P_R}{P_T} = G_T G_R \left(\frac{\lambda}{4\pi d}\right)^2 \left|1 + \rho e^{j\Delta\varphi} + (1-\rho)A e^{j\Delta\varphi} + \ldots\right|^2. \tag{2.70}$$

Both ρ and A depend on various factors, such as incidence angle, polarization, dielectric constants and frequency. For a particular case,

$$\rho = \frac{\sin\theta - K}{\sin\theta + K}, \tag{2.71}$$

in which θ is the incidence angle and K varies according to the previously mentioned parameters. For $\theta \simeq 0°$ (corresponding to the case in which the distance between the mobile terminal and the BS is much larger than the antenna heights) one can approximate $\rho \simeq -1$.

Besides that, ρ goes to -1 for frequencies above 100 MHz, and incidence angles smaller than $10°$. the terrestrial wave effects are perceived only for a few wavelengths above the surface, which simplifies the equation to

$$\begin{aligned}
\frac{P_R}{P_T} &= G_T G_R \left(\frac{\lambda}{4\pi d}\right)^2 |1 - e^{j\Delta\varphi}|^2 \\
&= G_T G_R \left(\frac{\lambda}{4\pi d}\right)^2 |1 - \cos\Delta\varphi - j\sin\Delta\varphi|^2 \\
&= G_T G_R \frac{2}{(4\pi d/\lambda)^2}(1 - \cos\Delta\varphi) \\
&= G_T G_R \frac{4}{(4\pi d/\lambda)^2}\sin^2\frac{\Delta\varphi}{2},
\end{aligned} \tag{2.72}$$

in which

$$\Delta\varphi = 2\pi\frac{\Delta d}{\lambda} \tag{2.73}$$

and $\Delta d = d_1 - d_0$

$$d_1 = \sqrt{(h_1 + h_2)^2 + d^2} \tag{2.74}$$

and

$$d_2 = \sqrt{(h_1 - h_2)^2 + d^2}. \tag{2.75}$$

The difference between the direct and the reflected paths distances Δd can be written as a function of h_1, h_2 e d. Then,

$$\Delta\varphi = \frac{2\pi d}{\lambda}\left\{\left[\left(\frac{h_1 + h_2}{d}\right)^2 + 1\right]^{\frac{1}{2}} - \left[\left(\frac{h_1 - h_2}{d}\right)^2 + 1\right]^{\frac{1}{2}}\right\}. \tag{2.76}$$

Using the approximation $(1+x)^{\frac{1}{2}} \simeq 1 + \frac{x}{2}$, for small values of x, one can write the previous equation as

$$\Delta\varphi = 4\pi \frac{h_1 h_2}{\lambda d}. \qquad (2.77)$$

The ration between the received and transmitted power can be written as

$$\frac{P_R}{P_T} = G_T G_R \frac{4}{(4\pi d/\lambda)^2} \sin^2 \frac{2\pi h_1 h_2}{\lambda d}. \qquad (2.78)$$

For small values of $\Delta\varphi$, $\sin(\frac{\Delta\varphi}{2}) \simeq \frac{\Delta\varphi}{2}$, then

$$\frac{P_R}{P_T} = G_T G_R \frac{4}{(4\pi d/\lambda)^2} \left(\frac{2\pi h_1 h_2}{\lambda d}\right)^2 \qquad (2.79)$$

$$\frac{P_R}{P_T} = G_T G_R \left(\frac{h_1 h_2}{d^2}\right)^2. \qquad (2.80)$$

This is the usual formula to compute the attenuation loss for cellular systems, which shows an attenuation that is superior to the one obtained from the free-space propagation model.

Introducing the transmit and receive antenna gains, the corresponding loss, in dB, is given by

$$L = -10 \log G_T - 10 \log G_R - 20 \log(h_1 h_2) + 40 \log d. \qquad (2.81)$$

From the previous equation, it is possible to derive two special relations, as follows:

$$\Delta P = 40 \log \frac{d_1}{d_2} \qquad \text{(Loss of 40 dB/dec)} \qquad (2.82)$$

$$\Delta G = 20 \log \frac{h_1'}{h_1} \qquad \text{(Gain due to elevation } - 6 \text{ dB/oct),} \qquad (2.83)$$

in which ΔP is the power difference, in dB, for distinct distances, and ΔG é is the gain, or loss, obtained from two different antenna heights in a cell.

For those measurements, the mobile antenna height gain is only 3 dB/oct, which is different from the gain of 6 dB/oct for h_1'. Therefore,

$$\Delta G' = 10 \log \frac{h_2'}{h_2} \qquad \text{(Antenna gain due to elevation } -3 \text{ dB/oct).} \qquad (2.84)$$

2.7.7 Propagation Over Water

The interference between the various multipaths, for distinct terrain conditions, has to be studied, and correction factors sometimes are introduced to account for those effects.

In general, the water relative permittivity ε_r, for a lake or for the sea are the same, but the conductivity is different. For the ocean, the permittivity, at a frequency of 800 MHz, is $\varepsilon_r = 80 - j84$. For a lake or river, the permittivity is $\varepsilon_r = 80 - j0.021$.

But, from the formula for the reflection coefficient, with a small incidence angle, both reflection coefficients, for the polarization of the horizontal and vertical waves are, approximately equal to one. If a phase change of $180°$ occurs, on the ground reflection point, the coefficient becomes minus one.

Figure 2.7 illustrates the BS and the mobile terminal, both above sea level. There is a direct wave and two reflection points. The formula can be used to find the strongest field intensity for the condition of propagation over water.

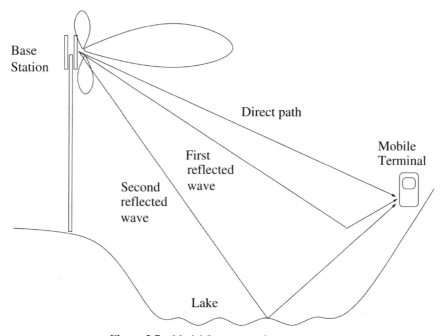

Figure 2.7   Model for propagation over water.

2.7.7.1 Fixed stations

The transmission between fixed stations over water, as exemplified in Figure 2.8, can be estimated as follows. The received power P_R is given by

$$P_R = P_T \left(\frac{1}{4\pi d/\lambda} \right)^2 \left| 1 + a_v e^{-j\phi_v} e^{j\Delta\phi} \right|^2, \tag{2.85}$$

in which P_T is the transmitted power, P_R is the received power, d is the distance between the BSs, λ is the wavelength a_v and ϕ_v are the complex amplitude and phase reflection coefficients, respectively, $\Delta\phi$ is the phase difference caused by the path difference Δd between the direct and reflected waves.

$$\Delta\phi = \beta\Delta d = \frac{2\pi}{\lambda}\Delta d. \tag{2.86}$$

The first part of Equation (2.85) is the formula for the free space propagation, which shows a decrease of 20 dB/decade, this is, this loss is obtained for a maximum distance of 10 km.

$$P_o = \frac{P_T}{(4\pi d/\lambda)^2}. \tag{2.87}$$

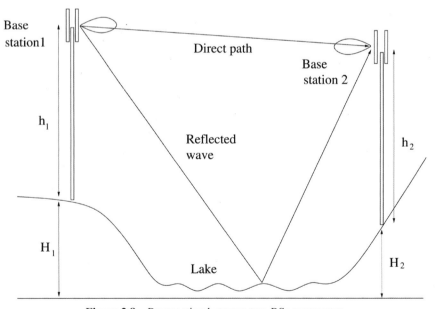

Figure 2.8   Propagation between two BSs over water.

The complex reflection coefficients $a_v e^{-j\phi_v}$ are computed using the following formula

$$a_v e^{-j\phi_v} = \frac{\varepsilon_r \sin\theta_1 - \left(\varepsilon_r - \cos^2\theta_1\right)^{\frac{1}{2}}}{\varepsilon_r \sin\theta_1 + \left(\varepsilon_r - \cos^2\theta_1\right)^{\frac{1}{2}}}. \tag{2.88}$$

If the incidence angle is small, $\theta \approx 0$, then $a_v \approx -1$ and $\phi_v = 0$.

The dielectric constant ε_r is different for distinct environments. But, when $a_v e^{-j\phi_v}$ is independent of ε_r, the reflection coefficient remains equal to -1, and it does not depend on the propagation over water, ice, dry or humid terrain.

The propagation between the fixed BSs is illustrated in Figure 2.8. Equation (2.85) becomes

$$\begin{aligned} P_R &= \frac{P_T}{(4\pi d/\lambda)^2} \left|1 - \cos\Delta\phi - j\sin\Delta\phi)\right|^2 \\ &= P_o \left(2 - 2\cos\Delta\phi\right), \end{aligned} \tag{2.89}$$

in which $\Delta\phi$ is a function of Δd, and can be obtained in the following.

The effective heights of antennas 1 and 2, above sea level, are, respectively,

$$h_1' = h_1 + H_1,$$
$$h_2' = h_2 + H_2,$$

as is shown in Figure 2.8, in which h_1 e h_2 are the actual antenna heights and H_1 and H_2 are the heights of the antenna towers above sea level.

In general, the antennas are positioned at a high altitude, and the reflection point over water is found halfway between the radiowave path. Difference Δd can be obtained from Figure 2.8, as

$$\Delta d = \sqrt{(h_1' + h_2')^2 + d^2} - \sqrt{(h_1' - h_2')^2 + d^2} \tag{2.90}$$

in which $d \geq h_1'$ and h_2', then

$$\Delta d \approx d \left[1 + \frac{(h_1' + h_2')^2}{2d^2} - 1 - \frac{(h_1' - h_2')^2}{2d^2}\right] = \frac{2h_1' h_2'}{d} \tag{2.91}$$

and Equation (2.86) becomes

$$\Delta\phi = \frac{2\pi}{\lambda} \frac{2h_1' h_2'}{d} = \frac{4\pi h_1' h_2'}{\lambda d}. \tag{2.92}$$

Examining Equation (2.89), one can observe the following conditions:

1. $P_R < P_o$. The received power is smaller than the power received due to propagation in free space, that is

$$2 - 2\cos\Delta\phi < 1 \quad \text{or} \quad \Delta\phi < \frac{\pi}{3}. \tag{2.93}$$

2. $P_R = 0$

$$2 - 2\cos\Delta\phi = 0 \quad \text{or} \quad \Delta\phi < \frac{\pi}{2}. \tag{2.94}$$

3. $P_R = P_o$

$$2 - 2\cos\Delta\phi = 1 \quad \text{or} \quad \Delta\phi = \pm\frac{\pi}{3}. \tag{2.95}$$

4. $P_R > P_o$

$$2 - 2\cos\Delta\phi > 1 \quad \text{or} \quad \frac{\pi}{3} < \Delta\phi < \frac{5\pi}{3}. \tag{2.96}$$

5. $P_R = 4P_o$

$$2 - 2\cos\Delta\phi = 4 \quad \text{or} \quad \Delta\phi = \pi. \tag{2.97}$$

2.7.7.2 Fixed and mobile stations

As described in Figure 2.7, there are two reflected waves, one over water the other near the mobile terminal, and a third one related to the direct path. The reflected wave, whose reflection point is located over water, is considered, because there is no object in the vicinity of this point, and there is a strong reflected energy.

The fractional reflected power, corresponding to the two reflected waves, reaches the mobile unit with no noticeable attenuation. The total received power is obtained as the sum of the three components.

$$P_R = P_T \left(\frac{1}{4\pi d/\lambda}\right)^2 \left|1 \quad e^{j\Delta\phi_1} - e^{j\Delta\phi_2}\right|^2 \tag{2.98}$$

in which $\Delta\phi_1$ and $\Delta\phi_2$ are the phase differences that corresponds to the path lengths differences between the direct and the two reflected waves, respectively. Usually, $\Delta\phi_1$ and $\Delta\phi_2$ are very small, therefore

$$P_R = P_T \left(\frac{1}{4\pi d/\lambda}\right)^2 \times |1 - \cos\Delta\phi_1 - \cos\Delta\phi_2$$
$$- j(\sin\Delta\phi_1 + \sin\Delta\phi_2)|^2. \tag{2.99}$$

Using the same analogy to approximate the propagation between a fixed station and a mobile terminal over water, one obtains

$$\cos\Delta\phi_1 \approx \cos\Delta\phi_2 \approx 1, \quad \sin\Delta\phi_1 \approx \Delta\phi_1 \quad \text{and} \quad \sin\Delta\phi_2 \approx \Delta\phi_2,$$

then

$$P_R = P_T \left(\frac{1}{4\pi d/\lambda} \right)^2 |1 - j(\Delta\phi_1 + \Delta\phi_2)|^2$$

$$= \frac{P_T}{(4\pi d/\lambda)^2} \left[1 - (\Delta\phi_1 + \Delta\phi_2)^2 \right]. \qquad (2.100)$$

In practical terms, $\Delta\phi_1 + \Delta\phi_2 < 1$, thus $(\Delta\phi_1 + \Delta\phi_2)^2 \ll 1$ and Equation (2.100) can be simplified to

$$P_R = \frac{P_T}{(4\pi d/\lambda)^2}. \qquad (2.101)$$

Equation (2.101) is the same formula obtained from the free space propagation condition. Therefore, one concludes that the propagation loss from the fixed to the mobile terminal, equivalent to 40 dB/decade, is different from the propagation over water. In this case, the it is possible to use the formula for free space propagation, or 20 dB/decade.

2.7.8 Foliage Loss

For this type of loss, several parameters and variations have to be taken into consideration. The type of leaf, branch, and trunk, the density of the trees and their distribution, and the tree height, relative to the antenna location are also considered.

In tropical zones, the vegetation is usually high and thick, and this difficults the signal penetration. In this case, the signal propagates from the top of the trees and is reflected towards the mobile device.

Experiments, for frequencies between 50 MHz and 800 MHz, and distances of 40 m to 4 km, in which the transmitter and receiver are surrounded by the forest, are presented in the following.

- The loss increases linearly in a logarithmic scale as the distance increases. An attenuation of 40 dB/decade, in the frequency of 800 MHz, and another one of 35 dB/decade, for 50 MHz, can be found.

Therefore, if the foliage loss is considered in the communication channel, at a frequency of 800 MHz, then a total loss is 60 dB/decade, which is equivalent to the sum of the free space loss, (20 dB/decade), plus the foliage loss, (20 dB/decade), and plus the the mobile environment loss (20 dB/decade) is expected.

The model is valid for the case of foliage along the propagation path. The foliage loss in a suburban area is 58.4 dB/decade.

- The loss increases exponentially with the frequency, in a logarithmic scale. At a distance of 4 km, the attenuation difference, between 80 and 800 MHz, is 20 dB, for vertical polarization, and 35 dB, for horizontal polarization.

The foliage loss is related to the frequency elevated to the fourth power (f^4), obtained from the Tamir prediction theory, increases as the horizontal polarization is lost (Lee, 1989).

- The difference in loss, as a result of the two polarization types ranges from 8 to 25 dB, for a frequency of 50 MHz e 1–2 dB for a frequency of 800 MHz. It is usual to obtain a smaller vertical polarization loss.
- The attenuation rate due to foliage, in the 50–800 MHz frequency range is:

$$0.005 - 0.3 \text{ dB/m} \quad \text{(horizontal polarization)};$$
$$0.005 - 0.51 \text{ dB/m} \quad \text{(vertical polarization)}.$$

- A delay of 0.2 µs is observed considering the following parameters: the receiver antenna is 8 m above the tree level, and the transmitting antenna is located in the middle of the trees, at a propagation distance of 160 m from the receiver.

2.8 Standard Predicting Models

This section presents the main prediction models for the electromagnetic field intensity, that are standardized by international commissions, such as the European Cooperation in Science and Technology (COST), an intergovernmental framework for cooperation in science and technology based in Brussels, Belgium, or used by mobile cellular communications operators.

2.8.1 Okumura Model

The Okumura model is a widely used empirical model for signal prediction in urban areas, that is applicable for frequencies in the range 150–1920 MHz, typically extrapolated up to 3000 MHz, and distances of 1–100 km.

The model produces a complete set of empirical data, for both natural and artificial structures, and can be used for BS antenna heights ranging from 30 to 1000 m. The information to build the model used data collected during measurement campaigns in the city of Tokyo, Japan (Okumura et al., 1968). Therefore, correction factors must be used to adjust the results to other cities.

The initial estimate comprehends a set of attenuation curves $A(f, d)$ as a function of the frequency f and distance d, for a BS antenna height of h_t of 200 m and a mobile terminal antenna height of 3 m,

The prediction curves are related to the free space loss. The correction factor G_{area} (gain), for a given terrain type, is also a function of the frequency. For different antenna heights, the following correction factors are obtained (Lee, 1992):

- Gain $G(h_t)$ of 6 dB per octave for the BS height

$$G(h_t) = 20 \log(h_t/200) \quad h_t > 10 \text{ m}.$$

Gain $G(h_r)$ of 3 dB per octave for the mobile terminal height

$$G(h_r) = 10 \log(h_r/3) \quad h_r < 3 \text{ m},$$
$$G(h_r) = 20 \log(h_r/3) \quad 3 \leq h_r \leq 10 \text{ m}.$$

Okumura proposed the following equation for the propagation loss

$$L = L_0 + A(f, d) - G_{area} - G(h_t) - G(h_r) \text{ (dB)}. \tag{2.102}$$

All the factors, that are available as a set of graphs, predicted by the model for different conditions, are expected to be adjusted in the course of the project. The Okumura model is not fitted for a computer, because of the various plots it requires. Therefore,

O model de Okumura não foi projetado para uso computacional, pois ele envolve várias curvas. Desta forma, Hata desenvolveu uma fórmula emprica para solucionar o problema.

2.8.2 Hata Model

Hata developed an empirical formula that can be programmed in a computer, an involves a set of analytical approximations for the propagation loss (Lee, 1992). The formula for the loss in urban areas is

$$L = 69.55 + 26.16 \log f - 13.82 \log h_t - A(h_r)$$
$$+ (44.9 - 6.55 \log h_t) \log d \text{(dB)}, \tag{2.103}$$

in which

$$150 \leq f \leq 1000 \text{ MHz},$$
$$30 \leq h_t \leq 300 \text{ m},$$
$$1 \leq d \leq 20 \text{ km}.$$

The correction factor $A(h_r)$ é is determined as:

- For medium-sized cities

$$A(h_r) = (1.1 \log f - 0.7)h_r - (1.56 \log f - 0.8)(\text{dB}),$$

in which $1 \leq h_r \leq 10$ m.

- For large cities

$$A(h_r) = 8.29 \log^2(1.54 h_r) - 1.1 \,(\text{dB}) \quad (f \leq 200 \text{ MHz}),$$

and

$$A(h_r) = 3.2 \log^2(11,75 h_r) - 4.97 \,(\text{dB}) \quad (f \geq 400 \text{ MHz}).$$

2.8.3 Lee Model

2.8.3.1 Loss prediction plot

The local averages are computed while the mobile moves along a given trajectory on the x axis. But, each local average is based on an x' related to the propagation path between the radiobase station and the mobile terminal at the corresponding point.

The computation of the loss curve is done on the x' axis. So, the local average, obtained from the signal mean stored along the x axis is changed to the x' axis.

A standard deviation of 8 dB is found as a function of several topographic contours, where the data is collected. The collected points, along the radioelectric propagation path, follow a lognormal probability distribution.

2.8.3.2 Area prediction model

This method uses the prediction of losses, but does not consider the terrain characteristics. If the propagation loss is obtained from the cartography of a mountainous terrain, this causes a discrepancy between the measured value and the corresponding value obtained from the prediction curve.

The area prediction requires the power at a certain distance from the transmitter, P_0, and the slope of the loss plot, β.

The received signal power is

$$P_R = \alpha_0 P_0 \left(\frac{r}{r_0}\right)^{-\beta} \left(\frac{f}{f_0}\right)^{-n}, \tag{2.104}$$

that can be expressed in dB as

$$P_R = \alpha_0 + P_0 - \beta \log \left(\frac{r}{r_0}\right) - n \log \left(\frac{f}{f_0}\right), \tag{2.105}$$

Table 2.2 Parameters for prediction losses in several propagation environments

Topography	P_0 (dBm)	β
Free space	-45	20
Open area	-49	43.5
Suburban area (USA)	-61.7	38.4
Urban area (Philadelphia)	-70	36.8
Urban area (Newark)	-64	43.1
Urban area (Tokyo)	-84	30.5

in which r is given in kilometers, and the distance where the measurements are taken is, usually, one mile, that is, r_0 equals 1.6 km. The correction factor α_0 is used to adjust for the difference in antenna height, power and gain, for the radiobase station and the mobile terminal.

The following conditions are assumed for the Lee model, and the parameters are listed in Table 2.2 (Lee, 1992):

- Frequency $f_0 = 900$ MHz
- Radiobase station antenna height (BS), 30.48 m.
- Transmitted power at the BS, 10 W.
- Radiobase station antenna gain, 6 dB, above the pilot gain.
- Mobile terminal antenna height, 3 m
- Mobile terminal antenna gain, 0 dB, above the pilot gain.

If the local conditions are different from the previous ones, then new parameters are computed, as in the following:

$$\alpha_1 = \left(\frac{\text{new BS antenna height (m)}}{30.48 \text{ m}} \right)^2, \tag{2.106}$$

$$\alpha_2 = \left(\frac{\text{new MT antenna height (m)}}{3 \text{ m}} \right)^\xi, \tag{2.107}$$

$$\alpha_3 = \left(\frac{\text{new transmission power}}{10 \text{ watts}} \right)^2, \tag{2.108}$$

$$\alpha_4 = \frac{\text{new BS antenna gain } (\lambda/2 \text{ of dipole})}{4}, \tag{2.109}$$

$$\alpha_5 = \text{different TM antenna correction factor.} \tag{2.110}$$

For the new parameters, the correction factor is

$$\alpha_0 = \alpha_1 \times \alpha_2 \times \alpha_3 \times \alpha_4 \times \alpha_5. \tag{2.111}$$

The value for n is found from empirical data, and is usually between 2 and 3, but the exact number is a function of the frequency and the topography of the area. For $f_0 < 450$ MHz, in suburban or open areas, the recommended value is $n = 2$. In urban areas, for $f_0 > 450$ MHz, it is recommended to use $n = 3$.

The value of ξ is also obtained from empirical data,

$$\xi = \begin{cases} 2, & \text{for the TM antenna height above 10 } m, \\ 3, & \text{for the TM antenna height below 3 } m. \end{cases}$$

2.8.3.3 General model

$$P_{\mathrm{R}} = -61.7 - 38.4 \log r - n \log \left(\frac{f}{900} \right) + \alpha_0 \quad \text{dBm (Suburban)}$$

$$= -70 - 36.8 \log r - n \log \left(\frac{f}{900} \right) + \alpha_0 \quad \text{dBm (Philadelphia)}$$

$$= 64 - 43.1 \log r - n \log \left(\frac{f}{900} \right) + \alpha_0 \quad \text{dBm (Newark)} \qquad (2.112)$$

$$\alpha_0 = 20 \log \left(\frac{h_1}{30.48} \right) + 10 \log \left(\frac{P_{\mathrm{T}}}{10w} \right) + (g_1 - 6) + g_2 + 10 \log \frac{h_2}{3.048}$$

$$= 20 \log h_1 + 10 \log P_{\mathrm{T}} + g_1 + g_2 + 10 \log h_2 - 64 \qquad (2.113)$$

in which the new values are P_{T}, given in watts, the BS and MT antenna height, h_1 and h_2, given in meters, the antenna gains are in dBd, in relation to the dipole gain, r is given in kilometers and f is given in MHz. The standard deviation is usually 8 dB.

2.8.4 COST231–Hata Model

This model is an extension of the Okumura–Hata model, and has the following constraints for the parameters: carrier frequency in the range $1500 \le f_c \le 2000$ MHz, radiobase station antenna height $30 \le h_b \le 200$ m, mobile terminal antenna height $1 \le h_m \le 10$ m (m), and distance range $1 \le d \le 20$ km.

The propagation loss, in dB, is given as

$$L_{\mathrm{p}} = A + B \log(d) + C, \qquad (2.114)$$

in which:

$$A = 46.3 + 33.9 \log(f_c) - 13.82 \log(h_b) - a(h_m), \tag{2.115}$$

$$B = 44.9 - 6.55 \log(h_b), \tag{2.116}$$

$$C = \begin{cases} 0, & \text{suburban area,} \\ 3, & \text{urban area.} \end{cases} \tag{2.117}$$

Although the Okumura–Hata the COST231–Hata models are limited to a radiobase station antenna height above 30 m, they can be used for antenna heights below that value, given that the surrounding buildings ceilings are below the average height of the antennas. But they are not recommended for distances shorter than 1 km, because the loss is dependent on the local topography.

2.8.5 COST231–Walfish-Ikegami Model

This model is used when the radiobase station antenna height is above, or below, the average height of the surrounding buildings. The loss, in dB, for line-of-sight (LOS) propagation is

$$L_P = 42.6 + 26 \log(d) + 20 \log(f_c), \quad d \geq 20 \text{ m.} \tag{2.118}$$

The model parameters are the distance d, in kilometers, and the carrier frequency, f_c, in MHz. The constant is determined for a distance of 20 m from the radiobase station, in order to consider free space propagation.

In case of non-line-of-sight propagation (NLOS), when the signal path is obstructed, the loss is expressed as a function of the buildings height, h_B, the street width w, the construction separation b, and the angle between the propagation direction and the position of the mobile terminal ϕ.

The propagation loss is composed of the following terms:

$$L_P = \begin{cases} L_o + L_K + L_H, & \text{for } L_K + L_H \geq 0, \\ L_o, & \text{para } L_K + L_H < 0 \end{cases} \tag{2.119}$$

in which L_o is the free space propagation loss, L_K is the knife-edge diffraction and L_H is the diffraction produced by multiple hard and rough surfaces.

The free space propagation loss is given by

$$L_o = 32.4 + 20 \log(d) + 20 \log(f_C). \tag{2.120}$$

The knife-edge diffraction loss is

$$L_K = -16.9 - 10 \log(w) + 10 \log(f_c) + 20 \log(\Delta h_M) + L_o, \tag{2.121}$$

in which

$$L_o = \begin{cases} -10 + 0.354\,(\phi), & 0 \le \phi \le 35° \\ 2,5 + 0.075\,(\phi - 35), & 35 < \phi \le 55° \\ 4.0 - 0.114\,(\phi - 55), & 55 < \phi \le 90° \end{cases} \tag{2.122}$$

and

$$\Delta h_{\mathrm{M}} = h_{\mathrm{B}} - h_{\mathrm{M}}.$$

The diffraction loss produced by multiple hard and rough surfaces is

$$L_{\mathrm{M}} = L_{\mathrm{S}} + k_{\mathrm{a}} + k_{\mathrm{d}} \log(d) + k_{\mathrm{f}} \log(f_{\mathrm{C}}) - 9 \log(b) \tag{2.123}$$

in which

$$L_{\mathrm{S}} = \begin{cases} -18 \log(1 + \Delta h_b) & h_b > h_{\mathrm{B}} \\ 0 & h_b \le h_{\mathrm{B}} \end{cases} \tag{2.124}$$

$$k_{\mathrm{a}} = \begin{cases} 54, & h_b > h_{\mathrm{B}} \\ 54 - 0.8\Delta h_b, & d \ge 0.5 \text{ km } e\ h_b \le h_{\mathrm{B}} \\ 54 - 0.8\Delta h_b d/0.5, & d < 0.5 \text{ km } e\ h_b \le h_{\mathrm{B}} \end{cases} \tag{2.125}$$

$$k_d = \begin{cases} 18, & h_b > h_B \\ 18 - 15\Delta h_b/h_B & h_b \le h_B \end{cases} \tag{2.126}$$

$$k_f = -4 + \begin{cases} 0,7(f_c/925 - 1), & \text{suburban area} \\ 1,5(f_c/925 - 1), & \text{urban area} \end{cases} \tag{2.127}$$

and

$$\Delta h_b = h_b - h_B.$$

The parameter k_a represents the incremental loss for radiobase station antennas that are below the surrounding buildings. The parameters k_d and k_f control the dependence of the multiple diffractions on the distance and frequency.

The model is useful, as long as the parameters are constrained to the following ranges: carrier frequency between $800 \le f_c \le 2000\,\text{MHz}$, radiobase station antenna height $4 \le h_b \le 50\,\text{m}$, mobile terminal antenna height $1 \le h_{\mathrm{m}} \le 3\,\text{m}$, and distance $0.02 \le d \le 5\,\text{km}$.

If no data is available about the local structures, roads and constructions, then the following values are recommended: $b = 20$–50 m, $w = b/2$, angle $\phi = 90°$, and $h_B = 3 \times$ number of stages plus ceiling of the buildings (m).

The COST231-WI model is more useful for $h_b \geq h_B$. For $h_b \approx h_B$ large prediction errors are expected, and for $h_b \ll h_B$, it is possible to obtain poor results, because some parameters are not considered in Formula 2.125.

2.8.6 Analytic Prediction Model for Urban and Suburban Environments

For environments that involve hard and rough surfaces, a basic knowledge of the propagation properties is necessary in order to create prediction models. The three main propagation processes are the following.

1. Free space propagation.
2. Knife-edge diffraction.
3. Diffraction from hard and rough multiple surfaces.

The total propagation loss L, in dB, is the addition of three independent terms: free space propagation, L_F, knife-edge diffraction, L_K, and Diffraction due to hard and rough multiple surfaces, L_H.

$$L = L_F + L_K + L_H. \tag{2.128}$$

The free space propagation loss can be written as a function of the wavelength, λ, and the distance between the transmitter and receiver, R.

$$L_F = -10 \log \left(\frac{\lambda}{4\pi R} \right)^2. \tag{2.129}$$

Using the Geometric Diffraction Theory (GDT), the knife-edge diffraction loss is given by

$$L_K = -10 \log \left[\frac{\lambda}{4\pi^2 r} \left(\frac{1}{\theta} - \frac{1}{2\pi + \theta} \right)^2 \right] \tag{2.130}$$

with

$$\theta = \tan^{-1} \left(\frac{\Delta hm}{x} \right)$$

$$r = \sqrt{(\Delta hm)^2 + x^2}$$

in which Δhm is the difference between the top of the obstacle and the mobile station antenna height, and x is the horizontal distance distance between the mobile station and the obstacle. A factor 2 is included to account for the influence of the knife-edge diffraction, and other diffractions, on the average signal power.

According to studies published by Xia and Bertoni, the diffraction resulting from hard and rough multiple surfaces can be written as

$$L_{\mathrm{H}} = -10 \log(Q_M^2). \tag{2.131}$$

The factor fQ_M can be written in terms of the Boersma function

$$Q_M = \sqrt{M} \sum_{q=0}^{\infty} \frac{1}{q!} \left(2g\sqrt{j\pi} \right)^q I_{m-1,q}, \tag{2.132}$$

for M diffractions. The dimensioning parameter g is given by

$$g = \Delta h_{\mathrm{b}} \frac{1}{\sqrt{\lambda d}} \tag{2.133}$$

in which Δh_{b} é is the height of the radiobase station antenna in relation to the average level of the obstacles, and d is the mean separation between the obstacles.

The Boersma function satisfies the recursive relation,

$$I_{M-1,q} = \frac{(M-1)(q-1)}{2M} I_{M-1,q} + \frac{1}{2\sqrt{\pi}M} \sum_{n=1}^{M-2} \frac{I_{n,q-1}}{\sqrt{M-1-n}}, \tag{2.134}$$

with initial terms

$$I_{M-1,0} = \frac{1}{M^{\frac{2}{3}}} \qquad e \qquad I_{M-1,1} = \frac{1}{4\sqrt{\pi}} \sum_{n=0}^{M-2} \frac{1}{n^{\frac{2}{3}}(M-n)^{\frac{2}{3}}}. \tag{2.135}$$

The decay index, γ, can be computed using the logarithmic derivative

$$s = -\frac{\log(Q_{M+1}/Q_M)}{\log[(M+1)/M]}, \tag{2.136}$$

to obtain,

$$\gamma = 2(1 + s). \tag{2.137}$$

2.8.6.1 Simplified model for near field

For urban and suburban areas that are composed of structures of uniform heights, it is convenient to deploy the antennas on the top of the buildings. This results in the following approximations $\Delta h_b = 0$ and $R = Md$.

Under this conditions, the multiple diffraction field Q_M is reduced to the simple solution

$$Q_M = \frac{1}{M} = \frac{d}{R}.$$ (2.138)

The simplified total loss is

$$L = -10\log\left(\frac{\lambda}{2\sqrt{2\pi}R}\right)^2 - 10\log\left[\frac{\lambda}{2\pi^2 r}\left(\frac{1}{\theta} - \frac{1}{2\pi + \theta}\right)^2\right] - 10\log\left(\frac{d}{R}\right)^2.$$

(2.139)

If the radiobase antenna is located inside the set of buildings, a factor of 2 is introduced in the free space term, to account for the refracted beams on obstacles surrounding the BS, which reinforce the signal in the direct path.

2.8.6.2 Simplified model: Antenna above the obstacle

To provide a larger coverage area, the BS antenna height is usually superior to the average height of the near buildings. Then, the multiple diffraction field reduces to

$$Q_M = 2.35\left(\frac{g}{M}\right)^{0.9} = 2.35\left(\frac{\Delta h_b}{R}\sqrt{\frac{d}{\lambda}}\right)^{0,9}.$$ (2.140)

The total loss is, then

$$L = -10\log\left(\frac{\lambda}{4\pi R}\right)^2 - 10\log\left[\frac{\lambda}{2\pi^2 r}\left(\frac{1}{\theta} - \frac{1}{2\pi + \theta}\right)^2\right]$$
$$- 10\log\left[(2.35)^2\left(\frac{\Delta h_b}{R}\sqrt{\frac{d}{\lambda}}\right)^{1,8}\right].$$ (2.141)

2.8.6.3 Simplified model: Antenna below the obstacle

Microcells employ antennas that are positioned below the average obstacle level. In this case, the propagation model can be separated into two cylindrical wave diffractions.

The cylindrical wave that is excited by a line below the average obstacle level is multiply diffracted, therefore the Geometrical Diffraction Theory can be used, resulting in the Xia and Bertoni multiple diffraction process (Xia, 1996).

If the incidence is on the top of the obstacle, at an angle

$$\phi = -\tan^{-1}\left(\frac{\Delta h_{\mathrm{b}}}{d}\right),$$

the field decreases due to the combined effect of the cylindrical wave diffraction process. This can be expressed in the following equation

$$Q_{\mathrm{M}} = \left[\frac{d}{2\pi\,(R-d)}\right]^2 \frac{\lambda}{\sqrt{(\Delta h_b)^2 + d^2}} \left(\frac{1}{\phi} - \frac{1}{2\pi + \phi}\right)^2. \qquad (2.142)$$

The total propagation loss is given by

$$L = -10\log\left(\frac{\lambda}{2\sqrt{2\pi R}}\right)^2 - 10\log\left[\frac{\lambda}{2\pi^2 r}\left(\frac{1}{\theta} - \frac{1}{2\pi + \theta}\right)^2\right] \qquad (2.143)$$

$$- 10\log\left\{\left[\frac{d}{2\pi\,(R-d)}\right]^2 \frac{\lambda}{\sqrt{(\Delta h_b)^2 + d^2}}\left(\frac{1}{\phi} - \frac{1}{2\pi + \phi}\right)^2\right\}^2$$

Again, because the antenna is in the middle of the buildings, a factor 2 is introduced in the free space propagation term, to take into account the beam that are refracted on the obstacles surrounding the BS, because this has the effect of reinforcing the signal in the direct path.

3

Cell Planning

3.1 Introduction

This chapter aims to introduce cellular mobile systems design and planning principles, which are not usually found in the common literature, even though it is part of the role of the engineer who works for a cellular phone service provider. The following sections introduce cell-planning concepts.

3.2 Network Planning

A network consists of a collection of communication links and nodes through which information is transported. Links can refer to the ones found in fixed telephone networks, cellular networks, or those found in satellite networks.

Any organization that makes use of such resources should be able to manage them effectively in order to supply high-standard services to their subscribers. By applying cell-planning techniques it is possible for a given organization to meet future connectivity demands using existing networks as a starting point.

Naturally, new developments are required as time passes by. Thus, cell planning is an active process that continuously uses the latest innovations, by regularly making some planning adjustments. Time is a key aspect in network planning.

Planning is necessary for large telecom companies as well as for privately owned companies, academic institutions, and governmental agencies.

It is possible to reduce network planning to an optimization problem in which, cost, reliability, public image, capacity, and potential growth are all important constraints. Therefore, the weight of each of those constraints influences the overall result, making it a non-trivial problem (Robertazzi, 1998).

This chapter focuses on cell planning, one of the most promising and interesting subareas of network planning.

3.3 Introduction to Cell Planning

With the growing demand for mobile communication services, the problems of optimal systems design and network planning are becoming more and more important.

In practice, many factors are taken into consideration in the design of mobile communication systems, such as, performance, capacity, cell coverage, required traffic, topography, and propagation characteristics.

Decisions concerning the number of cells and cell deployment, design parameters related to base stations (BSs) and mobile stations (MSs), and channel allocation should be taken at the same time. Cell deployment may be determined based on a particular cluster type, on coverage, traffic distribution and propagation environment. Design parameters related to BSs and MSs cannot be specified until cell allocation is complete. Finally, channel allocation may improve the performance of the system when it comes to quality of service and interference control, but it can only be determined after the cellular mobile network has been specified.

Cost is a key element in deciding upon the economic viability of any given communication system. A good design method should be able to optimize costs and consider other factors such as network performance criterion, traffic, and technology updates.

Many commercially available packages have been successfully applied to cellular mobile systems network planning. However, they do not include financial planning or cost factors in their planning approaches. On the other hand, some programs, such as the modeling system STEM, are merely decision tools for support to develop financial models and provide the analysis of the costs of cellular mobile networks, but they don't consider network planning in their cost models.

So as to fill this gap, cost and other network planning factors should be taken into consideration. That combination is of great interest to mobile networks service providers, who intend to develop an optimization method for planning systems that minimizes overall cost of the system and at the same time assures good system performance (Hao et al., 1997).

Operational research strategy, which entails planning based on hierarchical optimization, has been successfully used for the planning of large-scale manufacturing systems. In those cases, aggregated planning is not correctly devised as in complex systems it cannot be formulated or solved.

As such, due to the hierarchical nature of network planning the design of cellular mobile communication systems, a hierarchical optimization network planning approach is presented to determine the radio network architecture, or rather, the number and size of the cells, their allocation, gain parameters, and height of antennae and the transmitted power to the BSs and MSs.

Before that, in sequence, the main parameters that influence the mobile network specifications as well as the steps to be followed in the design, development and expansion processes of cellular systems are described.

3.4 Cell-Planning Methodology

The following topics show the required steps to be followed in order to plan and develop a cellular mobile system.

3.4.1 Planning Procedure

The planning process involves grouping data, such as, statistics, maps, service demand, and other relevant information, which are obtained by carrying out interviews and calculations. The following topics will define some essential data to plan a site.

3.4.1.1 Expansion needs

Expansion needs happen throughout a certain period of years and should be implemented in stages. That means that the expansion over a geographical area is covered within a given period of time. Some factors should be considered when defining those needs, such as:

- Service area or business area;
- Type of service;
- Population; and
- Area growth expectations.

3.4.1.2 Population density

There is some information that are essential for cell operation, such as, where traffic is located, who is responsible for the traffic, when traffic is specially high and the type of cell phones which are commonly used.

Population density provides the basis for the prediction of initial subscriber demands, including the expected number of subscribers in different areas. By calculating the number of potential subscribers in a given cell and their service rates, the number of necessary channels can be initially obtained.

3.4.1.3 Telephone tariff

It is usually obtained by means of a calculation based on the predicted service subscriber traffic, so as to define how many subscribers are necessary for the planning and where they are located in order to calculate usage rate and expansion rate.

It is possible to provide the number of subscribers according to a certain type of area (densely-populated urban area, urban area, suburban, rural, and others). The number of subscribers per area type is stored, as it is necessary for traffic analysis and radio network channeling.

3.4.1.4 Traffic demand

Based on the determined number of subscribers, traffic is measured in erlangs (the name of a Danish mathematician). An erlang can be defined by a circuit or a voice path busy for an hour. It is also the amount of time that a subscriber keeps the circuit busy either with voice or with data. Half that period of time is used for traffic demand prediction.

Traffic per subscriber is obtained by the average number of calls and the average call duration. Erlang per subscriber is usually measured at peak time. A table, called Erlang B, is used to estimate the number of organs (links, switches, routers, junctors, hubs, and such) necessary to sustain traffic at a given grade of service (GoS).

The formulation to calculate the amount of traffic offered by one subscriber is given by

$$A = \frac{NT}{3600},\tag{3.1}$$

in which T is the average call time in seconds, N is the number of times per hour that a subscriber places a call and A is the traffic offered by a subscriber in a system.

Given the amount of traffic offered by a subscriber, the total traffic for a system can be calculated by multiplying total traffic for a single user and the expected number of subscribers.

3.4.1.5 Quality of service

Grade of service is used to express the probability of a call being lost due to transmission loss or trunk congestion. GoS can be expressed in the percentage of lost calls (2% of lost calls) or in percentage of network reliability (98% of reliability).

3.4.1.6 Access types

The engineer needs to determine the type of service to be offered: analog, digital, or a combination of both. Based on that it is possible to determine the type of coverage to be offered and the necessary equipment.

3.4.1.7 Spectral allocation

Bandwidth and frequency band should be defined. Frequency constraints should also be defined for the area. Channel planning, operating frequency, and bandwidth are used to determine the best project.

3.4.1.8 Coverage

Before development stage, some calculations should be carried out to determine coverage area and coverage reliability. Some information can be provided or estimated based on previously collected data.

Building or operational preferences constraints may limit antenna height. There may also be regulatory requirements for a given type of coverage area. The height of the antenna also interferes in cell radius.

3.4.2 Planning Methodology

Once materials and data have been extracted from the planning process, it is possible to begin the development process.

Cell planning consists of three basic stages, which include cell planning, frequency planning, and overall system design. The main components in each stage are as follows:

- Cell planning;
- Area covered by each site;
- Antennae planning;
- Frequency planning; and
- Coverage and interference analysis.

The components are described as follows.

3.4.3 Cell Planning

3.4.3.1 Propagation models

Most engineers use a propagation tool for system modeling. The propagation tool provides a database that contains some information, such as terrain characteristics, morphology, demography, terrain boundaries, buildings, fading, roads, and site data.

Each tool calculates signal propagation loss at a site using a propagation model. Several propagation models are currently used in industry Okumura, Lee, Hata, Bullington, COST 231, Terrain Integrated Rough Earth Model (TIREM); or a common model based on others, depending on the tool supplier.

Once data is inserted into the propagation tool, it returns the necessary information for the analysis of an existing or a proposed cellular system. These tools allow the engineer to propose changes in a system before spending large amounts of money or wasting considerable time.

3.4.3.2 System in equilibrium

The coverage of a bidirectional radio communications system is decided based on the weakest transmission direction. The expressions uplink and downlink are used to refer to both link directions.

Transmission directions should be balanced in order to avoid interference, reduced system access, and extra costs. The equilibrium of the system can be calculated in order to assure that the strength of the uplink and downlink signals are approximately the same. Apart from that, the equilibrium of the system may also show whether the design is reasonable in relation to the level of available output power in all MSs. This equilibrium is in relation to the cell boundaries.

In order to calculate the equilibrium of the system between the BS and the mobile, some information is necessary such as BS receiver power and BS transmitter power; MS receiver power and MS transmitter power; BS and MSs antenna gain; propagation loss, and losses related to the hardware of the BS.

3.4.3.3 Determining the geographical area

The geographical area is divided into many regions in order to indicate densely populated urban area, urban area, suburban, and rural. Each region is selected based on terrain evaluation, traffic, and market characteristics.

3.4.3.4 Green field cell planning

Maps and terrain use information are equally important for cell planning. This planning is presented as a standard of hexagonal cells of different sizes and cell types obtained as described before. However, it may not be possible to use the site location for the BSs.

Other aspects should be considered before the site locations are defined, including radio environment in the area of the site location area. Obstacles situated in the direction of the radiating beams from the antenna, such as mountains or buildings, may cause shadow areas and temporal dispersion. Nominal cell planning should be used for orientation when the location of the site is chosen.

3.4.3.5 Areas covered by the site

Propagation studies and coverage requirements will probably define the location of the site. As it is almost impossible to initially choose the exact location of it, a general coverage area of the site is defined. The general coverage area of the site should consist of a circular area around the most suitable location for the site. The radius of the area covered by the site should be between 20 and 25% of cell radius. The proposed cell radius should start in the proposed covered area. Apart form that, the terrain elevation and the gradient of the area covered should agree.

3.4.3.6 Antenna planning

Many manufacturers produce antennae. They are produced to operate in specific coverage areas and conditions. The engineer should identify the type of the antenna to be used for the system development. At this stage, some specifications are necessary, defining physical structure, operating frequency, antenna radiating diagram, frequency band, antenna gain, and other technical information.

3.4.4 Frequency Planning

An important goal to be achieved in cellular system planning is the growth of traffic capacity. A great number of subscribers per square meter may use the system while maintaining an acceptable level of GoS and quality of conversation. Appropriate cell planning should assure that these objectives are met.

Cell planning, either good or bad, is the foundation on which a cellular system is established. Furthermore, planning should take into account not only initial traffic capacity cost, but also the cost of traffic growth throughout the following years.

As co-channel interference may pose a big problem, some effort to minimize such interference should be made. One way to deal with that is by frequency reuse, according to the minimum reuse distance.

Frequency reuse in a cell of a given size, depends on the number of frequency groups in the channel reuse model. The greater the number of frequency groups, the greater the distance.

Cell clusters (for example, a group of seven cells per cluster for a 7/21 plain) may have frequencies established in a model that reduces the carrier-to-interference ratio (C/I). The frequency determination technique Knight's Move is used to determine the location of a group of frequencies or channels and the location of the nearest hexagons containing the same channels.

3.4.4.1 Planning for the future

It is important to plan system growth from the beginning. The following items may be necessary to be included in the company growth plans:

- Transition from omnidirectional sites to sectorized ones.
- Cell partitioning methods.
- Spectral expansion availability.
- Technology updates.

3.4.4.2 Transition from omnidirectional to sectorized

The transition from omnidirectional to sectorized is necessary when the rise in congestion shows that better interference control and greater number of channels are necessary. Although omnidirectional sites have greater traffic capacity than sectorized ones, the transition from omnidirectional to sectorized ones is necessary in order to reconfigure them.

Due to system interruption during the transitional period, it is better to go through with the transition faster so as not to disturb the provision of service.

The co-channel interference problem in an omnidirectional system is most noticeable when the MS is on the border of the cell, moving toward the cell with which the cell interferes. Sectorizing allows interference reduction by means of an additional gain between 6 and 15 dB in relation to the front/back of the antenna.

3.4.4.3 Interference and coverage analysis

Among the different roles of the radio engineer there are some tests, such as the one to create a system with optimized coverage, with minimum interference. By using a propagation calculation tool, the engineer is able to model the system and view the coverage area and interference area.

Coverage is based on the strength of the signal that is required at a given site. For example:

Signal Strength (dBm)	Description
−85	Excellent coverage
−95	Good coverage
−105	Bad coverage

The typical RF interference margins that affect the implementation and realization of a system are:

- signal-to-noise ratio (SNR; related to the sensitivity of the receiver and soil noise);
- co-channel interference (C/I);
- adjacent channel interference (C/A); and
- bit error rate (BER).

The following section presents some aspects related to the expansion of the cellular system.

3.5 Expansion of the Cellular System

Like other mobile communication services, cellular systems have experienced a growing demand for new subscribers lately. In contrast to greenfield cell planning, which is both very fast and modular, expansion planning demands a detailed study of the network behavior and its users.

There are several ways to expand a system. In order to choose the best method, some factors should be taken into consideration, such as, the current growth and development stage of the service, availability of resources and mainly the service demand.

The growth itself can be in relation to coverage or traffic, processing capacity, and spectral efficiency.

3.5.1 Coverage Area and Traffic

The capacity of a cellular system is mainly related to its ability to handle a great number of subscribers, keeping the quality of the conversation. Cellular system capacity is directly proportional to the MSC, as a rise in the number of mobile subscribers implies a rise in the central processing load.

A rise in the number of subscribers also implies reduction in cell sizes, with transmitter power reduction and tower height reduction, which may raise the number of handoffs per cell and, consequently, demand more processing capacity.

3.5.1.1 Traffic

Traffic handling capacity of a cellular system is expressed erlang/km^2 and depends on the number of available channels, cell size, and the frequency reuse model used.

The smaller the cell, the greater the system traffic handling capacity. There is, however, a minimum distance to be considered between cells, whose limit is given by the co-channel interference levels.

3.5.1.2 Traffic channel allocation

Pedestrian and traffic density in a given covered area is a critical parameter and as such should be determined before system design. Traffic records in the rush hour may be restricted to different zones within a service area. The choice of initial BSs sites should be based on signal coverage in heavy vehicular or pedestrian traffic. That means that the best place to place the BSs is in the center of those zones.

3.5.1.3 Computation of traffic capacity

The number of MSs that may be reached and the number of voice channels are important parameters for the planning of a given cell. Such planning should take into account the desired GoS, that is, the allowed percentage of calls which are not completed due to congestion.

For a given GoS, the relation between the number of channels and maximum traffic density per channel follows the Erlang set of curves, as shown in Figure 3.1.

3.5.1.4 Traffic unit

The traffic unit, erlang or Erl, is defined as the average number of concurrent calls for a given period of time. The unit is named after Agner Krarup Erlang (1878–1929), a Danish mathematician who was a pioneer of Telephone Traffic Theory.

For organs that hold ordinary calls, the period is usually an hour (Rouault, 1976). Traffic A, in erlangs, is usually obtained by (Alencar, 2000)

$$A = \lambda t_m, \qquad (3.2)$$

in which λ = number of calls per unit of time, or the number of new occupations per unit of time and t_m = the average duration of occupations expressed in the same unit of time.

It is also possible to use other traffic measurement methods. Suppose that for a given period of time T, N occupations take place. The occupation of a number n of organs lasted t_n. The duration of the occupations that were already

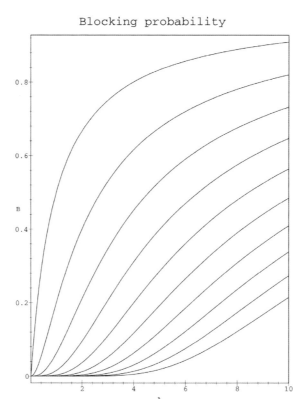

Figure 3.1 Erlang set of curves for the blocking probability.

in progress during the time interval T is computed from $t = 0$. Likewise, the duration of the occupations that are still in progress at $t = T$ is computed up to that moment (Siemens, 1975).

Traffic is obtained by

$$A = \frac{1}{T} \sum_{n=1}^{N} t_n. \tag{3.3}$$

Figure 3.2 shows the different stages to be followed during the design of a mobile communication system (Yacoub, 1993a).

3.5.1.5 Cell coverage

The system should serve the largest possible area, but by using radio signals, it is not always possible to cover 100% of the desired area. That is due to irregularities in the terrain boundary. An attempt to obtain full coverage may result in having to deal with the new constraints, cost and interference.

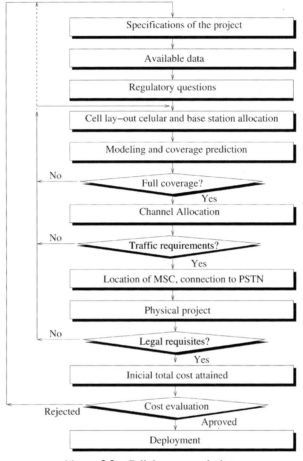

Figure 3.2   Cellular system design.

Thus, it is possible to raise the transmitted power to better cover points with low signal strength, or rather, insufficient reception. However, it will mean significant additional cost. Apart from that, the greater the transmitted power, the harder it is to control interference.

When radio signal quality and range are growing concerns for satisfactory system operation, some alternative solutions may be found to the system coverage problem. Some such techniques include inserting new BSs, raising antenna output power with proper interference control, eliminating shadow areas and superimposing cells.

Some procedures shall be followed when there are any changes in the configuration of cell coverage, such as, frequency plan and power control revision, so that co-channel interference and adjacent channel interference level are within acceptable levels. It is also compulsory to revise neighboring and adjacent cell allocation plans, which are essential for handoff and call forwarding mechanisms.

In general, it is possible to select cell type in any given situation, trying to adapt system parameters to the particular features of each region; for example, in urban areas where extra capacity is needed to accommodate visitors, coverage may be obtained by means of micro- or picocells, with radius in the range of dozens or hundreds of meters, respectively.

3.5.1.6 Traffic capacity

Nowadays, the main goal of service providers is to raise traffic flow capacity and reduce congestion in cells with a large number of users, such as the ones in city centers. Systems with little or no capacity to handle traffic and with high congestion mean low turnover, which is surely not what service providers have in mind.

The frequency spectrum is a limited resource, so in order to serve higher traffic flow, the only option is to improve the efficiency of the use of the available frequencies. That being the case, traffic growth is followed by the reduction of BSs transmitted power, by the use of antennae in lower points and/or internal points and by the growth of shadow areas, though the latter is not intended or planned.

Supplementary coverage, provided by cells in a higher hierarchy, is used to eliminate shadow areas. Those particular cells are identified by external antennae in higher places and high power use and that type of arrangement is known as the umbrella model.

3.5.2 Spectral Efficiency

One way to raise radio system capacity is to improve access. Therefore, multiple access techniques, such as, frequency division multiple access (FDMA), initially applied to analog systems, as well as time division multiple access (TDMA) and code division multiple access (CDMA), introduced by digital systems, have been used. As some digital systems operate in the same frequency range of analog systems, such as cellular systems, they are considered replacement technologies for analog ones.

However, it is not possible to waste the money invested in systems which are already in operation. Thus, the transition from one cellular system generation to the other should be softly carried out.

Frequency band allocation for the use of telecommunication services is a subject that is commonly raised in international meetings related to radiocommunication systems. In that respect, the systems that arrived first had some advantage over the latest ones.

Nowadays, however, frequency allocation for a given service is previously established by standardizing organizations, such as ITU (International Telecommunications Union).

Technically speaking, splitting the spectrum into smaller frequency bands reduces its efficiency, but two factors should be considered. One of them is service diversity, which has to be assured so as to provide competitiveness among service providers and boost the development of systems with higher capacity and better quality of service; the other is related to avoiding a huge number of systems operating in different frequency bands in other countries, hindering the possibility of global services.

Digital multiple access techniques have considerably improved wireless communications, but the transition from analog systems to digital systems should be accompanied by the following factors: using multiple access techniques to raise system capacity, cutting transition costs, better quality of services, greater diversity of services, and transparency guarantee for the final user. When it comes to the quality of services, special attention should be given to the users of the current system so that they do not experience any degradation in the GoS, once part of the allotted band for that system has been removed as part of the process of the deployment of the new cellular system. Surely, the subscribers that already own terminals will not be willing to waste their investment or buy a new terminal, unless they're interested in services that are not provided by the old system. As for the service providers, they would not also be willing to exchange their subscribers' old terminals just for the sake of deploying a new platform.

3.5.3 Processing Capacity

A rise in the number of users and their mobile profiles directly impact on the processing load in the Main Switching and control Center (MSC). The moment it reaches critical values, new processors, or new control centers should be added in order to handle the rise in the traffic experienced by the system. Therefore, multiple processing centers may coexist in the system.

If that is the case, it is important to wisely allocate not only the coverage areas and their access sites, but also the traffic associated to each site. Otherwise, there may be chaos in inter-system signaling or unbalanced processing load in the system, with overloaded processing centers and underused ones.

Excessive signaling jeopardizes the quality of services as well as reduces calling processing capacity, as part of its capacity is reserved for signaling. In some situations the percentage of reserved capacity is 30%. Signaling in-between systems is caused by several factors, such as, log updates, roaming and handoffs, especially the handoffs that involve BSs in different control areas. A handoff takes place when a MS moves from the area covered by one BS to the area covered by another BS changing the voice channel during conversation. If the BSs belong to different MSCs, the handoff is said to be in-between systems, or in-between centers.

3.5.3.1 Cellular system partitioning

Handoff is devised to provide mobility to cellular system users. However, the price paid for the facility affects the complexity and, consequently, the cost of the service. As user mobility is directly related to handoff, it defines how control areas should be partitioned between MSCs, so as to minimize handoffs between different systems. Only the BSs that hold heavy user traffic between them, are left under control of the same MSC.

In order to effectively partition a cellular network, it is necessary to develop a model that is able to estimate the costs of possible system partitions. One such model is described below.

3.6 Model to Estimate Costs

Mobile systems planning can be seen, mathematically, as a complex combinatorial optimization problem. Optimization algorithms like branch-and-bound or dynamic programing cannot be used in this situation, as they cannot yield the optimal solution in reasonable time. Therefore, the model proposed by Hao et al. (1997) is described as follows. It is based on cost estimates and from which an alternative optimization method is derived to determine adequate number of cells and the best locations of BSs.

3.6.1 Problem Characterization

In order to design a mobile communication system to serve a given study area, the whole area is initially divided according to a region classification, such as, urban, suburban, and rural. Uniform traffic distribution is initially assumed, with traffic peak usually happening in the city center and local peaks being confined to suburban centers. The covered area probability related to specified coverage, P_a, is provided.

The location probability of an MS near the cell boundary P_l and the threshold level of the received signal at the boundary region P_{cell} can be obtained from the coverage probability P_a and the required signal quality, namely the signal-to-noise ratio (SNR). The GoS is given by the blocking probability P_{block} for placing calls during the busiest hour (BH).

The optimal architecture is defined by determining the number and size of cells, the location of each BS and the parameters for both the BSs and the MSs, which can ensure the required system performance and minimize total system cost. The cost of the BS is obtained by the cost of hardware and installation, antennae, buildings and towers, as well as transmitters and receivers.

Many factors should be considered in the system design and many decisions should be made in different layers as the design progresses. The most important factors are: specification of system performance, cell coverage, traffic distribution, topography, propagation data, and system complexity. Since many decisions are involved in the process, the network planning process itself is considered to be hierarchical.

3.6.2 Design and Network Planning

A hierarchical optimization planning method (HOP) for the design of mobile communication networks is presented as follows. The three-level structure of the network planning as proposed by Hao et al. (1997) is shown in Figure 3.3.

At the first level, the upper bound of the number of BSs to meet the demand of the region and the cell ranges are defined. Traffic required at BH, coverage requirement and terrain features in study area are input parameters to the HOP. Propagation parameters are chosen for ordinary conditions. The goal is to cover the study area and meet the average traffic demand with minimum number of sites.

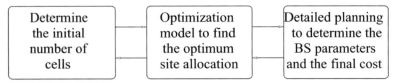

Figure 3.3   Hierarchical structure of the network planning as proposed by Hao et al. (1997).

At the second level, a combinatorial optimization procedure is applied to yield the number of cells and the best BSs locations. The main goal is to minimize overall system costs and at the same time guarantee excellent coverage, while keeping non-uniform traffic demand. The approach also considers different subscriber traffic densities and the particular terrain features in different service areas. As mentioned before, the study area, initially classified as rural, suburban and urban, is later subdivided into grids. Environment information, subscriber density and average elevation data for each grid can be obtained from a geographical information system (GIS).

At the third level, specific cell parameters, such as standard, antenna gain, transmitted power, antenna height, and channel use are defined. Thus, the overall cost is estimated.

As planning is concerned, system performance is highly dependent on the decisions made at each level. Furthermore, decisions made at one level shall impact on the other levels as seen on Figure 3.3.

In order to apply the model, the following relationships must be considered: the relationship between coverage and demand, cell range, and signal strength at cell boundary; the relationship between propagation loss, physical structures, and terrain characteristics; and the relationship between equipment and cost.

Cellular mobile service providers regularly apply Hata's propagation model to estimate transmission loss. The model's acceptance is due to the fact that it defines signal propagation in different types of environments, such as, urban, suburban, rural with smooth, or irregular terrain. Hata's propagation formula for urban areas is given by

$$
\begin{aligned}
L_{\mathrm{u}}(\mathrm{dB}) = {} & 69.55 + 26.66 \log(f) - 13.82 \log(h_{\mathrm{b}}) - a(h_{\mathrm{m}}) \\
& + [44.9 - 6.55 \log(h_{\mathrm{b}})] \log(d)
\end{aligned} \tag{3.4}
$$

in which the correction factor for the antenna height of MSs $a(h_{\mathrm{m}})$ is

$$
a(h_{\mathrm{m}}) = [1.1 \log(f) - 0.7]h_{\mathrm{m}} - [1.56 \log(f) - 0.8] \tag{3.5}
$$

for medium-sized and small-sized cities, and

$$
a(h_{\mathrm{m}}) = 3.2[\log(11.75\, h_{\mathrm{m}})]^2 - 4.97 \tag{3.6}
$$

for large cities, but for frequencies $f \leq 400$ MHz.

Propagation losses for suburban and rural areas L_{su} e L_{rur} are shown in Equations (3.7) and (3.8).

$$L_{\text{su}} = L_{\text{u}} - 2[\log(f/28)]^2 - 5.4, \tag{3.7}$$

$$L_{\text{rur}} = L_{\text{u}} - 4.78[\log(f)]^2 + 18.33\log(f) - 35.94. \tag{3.8}$$

Hata's formulas are restricted to the frequency range 150–1000 MHz, antenna height of the BS between 30 and 200 m, antenna height of the MS between 1 and 10 m and distance d from 1 to 20 km.

The HOP as presented by Hao et al. (1997) is described in detail in the following section.

3.6.3 Green Field Dimensioning

The minimum number of sites is initially determined based on coverage and traffic demand in the study area. The worst-case scenario is used for the calculation, so as to provide an upper bound to the number of cells in the system. A reuse factor of cells of $k = 7$ is used. The probability of covered area P_{a} and the blocking rate for subscribers P_{block} are given. Mobile traffic demand in a given covered area is the overall number of originating calls from the MS at BH and prediction is based on vehicular traffic density in the coverage area. Given a number of originating calls, traffic in an area is converted into de number of subscribers at BH in the area.

The following parameters are defined:

C_{pt} Traffic that each cell can handle (number of subscribers/hour) based on the number of channels in each cell and the blockingrate P_{block}

C_{tt} Overall traffic in the service area under study (number of subscribers/hour)

P_{cell} Threshold level of received signal strength at the cell boundary

P_{pp} RF output pcak power (dBW)

P_{t} Transmitting antenna input power (dBW)

P_{r} Receiving antenna received power (dBW)

$g_{\text{b}}, g_{\text{m}}$ Antenna gains for the BS and MS, respectively (dB)

$h_{\text{b}}, h_{\text{m}}$ Antenna heights for BS and MS, respectively (m)

d Average cell radius (km)

S Overall area of service system (km^2).

Coverage performance is initially evaluated. The loss in the link between the transmitter and receiver is given by Equations (3.9) and (3.10).

$$P_{\text{r}} = P_{\text{t}} + g_{\text{b}} - L(d) + g_{\text{m}}, \tag{3.9}$$

$$P_{\text{t}} = P_{\text{pp}} - l, \tag{3.10}$$

in which $L(d)$ is the transmission loss (dB) and l is the combined loss in the insulator, combiner and radio frequency (RF) cables.

The upper bound in the number of cells in the whole service area is given by Hata's propagation model for urban area. The models for suburban and rural areas are considered at the next planning level.

The following conditions are considered, e.g., $P_{pp} = 10$ W, $h_b = 30$ m, $h_m = 3$ m, $g_b = 12$ dBi, $g_m = 2$ dBi, $l = 4$ dB, and $f = 900$ MHz. Accordingly, Equation (3.5) for propagation loss L becomes

$$L(d) = 123.73 + 35.22 \log(d). \tag{3.11}$$

So as to guarantee coverage demand,

$$P_r = -73.73 - 35.22 \log(d) \geq P_{cell} \tag{3.12}$$

i.e.,

$$\log(d) \leq \log(d_{max}) = (-P_{cell} - 73.73)/35.22 \tag{3.13}$$

in which d_{max} is maximum cell radius in an urban area in a big city.

Thus, the minimum number of cells is given by

$$n_1 = S/\pi d_{max}^2. \tag{3.14}$$

If traffic pattern is more important than coverage pattern, the number of cells is determined based on traffic and the minimum number of cells is given by Equation (3.15)

$$n_2 = C_{tt}/C_{pt}. \tag{3.15}$$

The minimum number of cells is

$$n = \max\{n_1, n_2\}. \tag{3.16}$$

The minimum number of cells n as obtained in Equation (3.16) is the upper bound to be considered in order to obtain a cost-effective design and is used to determine the cell average radius $d = \sqrt{\left(\frac{S}{n\pi}\right)}$.

3.6.4 Optimum Allocation and Cell Dimensioning

Non-uniform traffic distribution in the study area is assumed at this level. Data referring to terrain structure and environment, traffic density and terrain elevation is stored for each grid. As the upper bound of the number of cells is already known from the previous level, it is possible to define which grid belongs to which cell.

Thus, the number of cells and size of each cell as well as cell allocations are determined. It is important to point out that a cell consists of many neighboring grids belonging to the same region. A combinatorial optimization model is applied to determine which grids belong to which cell as well as the optimal parameter values for each BS. In the model, a weighted cost function is used with coverage and non-uniform traffic demand as constraints. The goal of the model is to minimize overall system cost. The number of cells can be even reduced if light traffic is considered.

3.6.4.1 Mathematical modeling of the economic optimization

The following decision variables are introduced:

$$x_{ik} = \begin{cases} 1, & \text{if grid } i \text{ belongs to cell } k; \\ 0, & \text{if grid } i \text{ doesn't belong to cell } k. \end{cases}$$

$$Y_k = \begin{cases} 1, & \text{if there is a grid in cell } k; \\ 0, & \text{if cell } k \text{ is empty, i.e., with no grids.} \end{cases}$$

(3.17)

The parameters are defined as follows:

$$G_{i1} = \begin{cases} 1, & \text{grid } i \text{ belongs to urban area;} \\ 2, & \text{grid } i \text{ belongs to suburban area;} \\ 3, & \text{lot } i \text{ belongs to rural area.} \end{cases}$$

G_{i2} Traffic density in grid i (number of subscribers/hour);

n Total number of cells;

m Total number of grids;

C_{so} Fixed cost of switching operations, hardware and installation;

C_{cell} Cost of hardware and BS installation;

C_a Cost coefficient for antenna in relation to its gain;

C_t Cost coefficient for transmitter and receiver in relation to its transmitted power;

P_{tk} BS transmitting power in cell k, in which $P_{LB} \leq P_{tk} \leq P_{UB}$, in which P_{LB} e P_{UB} are the lower and upper bounds, respectively.

g_{bk}, g_{mk} BS and MS antenna gains in cell k, in which $g_{LB} \leq g_{bk} \leq g_{UB}$ in which g_{LB} e g_{UB} are the lower and upper bounds, respectively.

h_{bk}, h_{mk} BS and MS antenna heights in cell k, respectively.

d_k Radius of cell k.

S_g Grid area.

The Economic Optimization Model (EOM) for the mobile communication model is defined as follows. EOM:

$$\min f_{\mathrm{c}}(s) = C_{\mathrm{so}} + \sum_{k} Y_k (C_{\mathrm{cell}} + C_{\mathrm{a}} g_{bk} + C_t P_{tk}) \tag{3.18}$$

subject to the following constraints:

$$P_{tk} + g_{bk} + g_{mk} - L_k(d_k) \geq P_{\mathrm{cell}}, \quad k = 1, 2, \ldots, n \tag{3.19}$$

$$\sum_{i} x_{ik} \cdot G_{i2} \leq C_{\mathrm{pt}}, \quad k = 1, 2, \ldots, n \tag{3.20}$$

$$x_{ik} \cdot x_{i'k} \cdot (G_{i1} - G_{i'1}) = 0, \quad k = 1, 2, \ldots, n, i, i' = 1, 2, \ldots, m, \; i \neq i' \tag{3.21}$$

$$\sum_{k} x_{ik} = 1, \quad i = 1, 2, \ldots, m \tag{3.22}$$

$$\pi d_k^2 = \sum_{i} (x_{ik} \cdot S_{\mathrm{g}}), \quad k = 1, 2, \ldots, n \tag{3.23}$$

$$d_k(Y_k - 1) \geq 0, \quad k = 1, 2, \ldots, n \tag{3.24}$$

$$d_k > Y_k - 1, \quad k = 1, 2, \ldots, n \tag{3.25}$$

$$x_{ik}, Y_k \text{ is either 1 or 0 for every } i \text{ and } k. \tag{3.26}$$

In the EOM model, the goal of the cost function $f_{\mathrm{c}}(s)$ is to minimize overall system cost. Condition 3.19 guarantees coverage performance, while Condition (3.20) guarantees non-uniform traffic demand.

Conditions (3.21)–(3.23) assure that neighboring grids belonging to the same region (urban, suburban and rural) are part of the same cell. The relations between d_k and Y_k, i.e., $Y_k = 0$ for $d_k = 0$ and $Y_k = 1$ for $d_k > 0$, are given by Conditions (3.24)–(3.26).

The signal propagation loss $L_k(d_k)$ is calculated using Hata's prediction model. From Equations (3.5) and (3.14), considering $P_{\mathrm{pp}} = 10\,\mathrm{W}$, $h_{\mathrm{b}} = 30\,\mathrm{m}$, $h_{\mathrm{m}} = 3\,\mathrm{m}$, $g_{\mathrm{b}} = 12\,\mathrm{dBi}$, $g_{\mathrm{m}} = 2\,\mathrm{dBi}$, $l = 4\,\mathrm{dB}$, e $f = 900\,\mathrm{MHz}$, $L_k(d_k)$ is given by

$$L_k(d_k) = \alpha + 35.22 \log(d_k) \tag{3.27}$$

in which $\alpha = 123.73, 113.79$ and 102.22 for a cell k covering urban, suburban, and rural (almost open) areas, respectively.

3.6.4.2 SA to solve the EOM model

The Economic Optimization Model (EOM) Model is a complex optimization problem with many variables and hard constraints. Thus, no classical optimization algorithm can be used to obtain an optimal solution in reasonable time. Therefore, Hao et al. (1997) has developed an algorithm based on SA to solve the problem and has been able to obtain near-optimal solutions with low computational effort.

Simulated annealing is a method commonly used for combinatorial problems to obtain approximate solutions. It has been successfully applied in several fields, such as computer design, image processing and channel assignment. The algorithm is an iterative technique that mimics the physical process of heating a solid up to its melting point, subsequently, cooling it down until it crystallizes in its minimal energy state.

The following aspects should be considered when applying the SA algorithm in order to solve an EOM problem: configuration space, cost function, and neighborhood structure.

a. Configuration space – It is considered the set of all possible solutions x_{ik}, Yk that satisfy the coverage constraint given by Equation (3.19) and Constraints (3.21)–(3.26).

b. Cost function – In system design, coverage performance is initially considered. For a given number of cells, very few possible solutions can be found if both coverage and traffic constraints are considered, as non-uniform traffic distribution is assumed. Therefore, by inserting the traffic constraint (3.20) in the cost function so as to minimize overall equipment costs for BSs, it becomes

$$f_c(s) = \sum_k Y_k \cdot \left\{ C_{cell} + C_a g_{bk} + C_t P_{tk} + C_{traf} \left[\sum_i x_{ik} G_{i2} - C_{pt} \right]^+ \right\}$$
(3.28)

in which the function $[x]^+ = \max(0, x)$.

c. Neighborhood structure – The neighborhood of a solution s, $N(s)$, is obtained by transitioning grid k from its present cell i to a neighboring cell j, if Conditions (3.21)–(3.23) are satisfied.

In order to apply SA to the problem, the initial temperature, given by the ratio between accepted transitions and proposed transitions initial temperature should be defined. Then, a temperature decrement rule based on the standard deviation of the cost function is derived. It is given by

$$t_{k+1} = t_k \exp(t_k \Delta_c / \sigma_k^2) = t_k \exp(-\lambda t_k / \sigma_k)$$
(3.29)

in which σ_k is the standard deviation of the cost function at temperature t_k and Δ_c is the decrease in the average cost at two subsequent temperatures t_k and t_{k+1}.

The final temperature is reached when the difference between the maximum and minimum costs at a given temperature is equal to the maximum difference in the cost of an accepted transition at that same temperature.

3.6.5 Planning the Cellular Architecture

In the cellular system planning and design processes, the initial planning of a new network or the conversion from an existing analog network into a digital one involves the following aspects: capacity, coverage, cost, voice quality, quality of service, and further growth.

Cellular network planning serves as a guideline for the installation of new BSs, or a change in the current active ones. Cellular network engineering entails BS Network design concepts, such as, topographical study of the serviced area, field research, frequency spectrum use, transmission network planning, system performance prediction, and final system adjustment (Qualcomm, 1992).

The aforementioned parameters are obtained based on traffic demand. The traffic to be served can be calculated from the Busiest Hour (BH), the population distribution, the spatial distribution of automobiles and income distribution.

After deployment, the traffic handling capacity of a cellular mobile system, in erlangs per square kilometer or Equivalent Telephone Erlangs (ETEs) per square kilometer, may be increased by means of smaller cells and frequency reuse.

Effective planning for the deployment of the concept of universal personal communications involves the use of different cell structures, according to the operational environment. The cell structures to handle the Universal Mobile Telecommunications System (UMTS) network range from conventional macrocells to in-building picocells. That is shown in Table 3.1.

Table 3.1 Cellular structures to handle UMTS

Type of Cell	Coverage (m)	Power (W)	Antenna Height
Acrocell	>1.000	1–10	>30
Microcell	<1.000	0.1–1	<10
Picocell	5–30	0.01–0.1	Ceiling
Umbrella Cell	>1.000	1–10	>30
Road cell	100–1.000	<1	<10

Macrocells are used in conventional cellular radio systems. Coverage is maximized to reduce infrastructure costs. In UMTS macrocells are used to provide coverage for areas with low terminal density. Macrocells have an RMS delay spread ranging from 0.1 to 10 µs, which allows maximum transmission rate of 0.3 Mbits/s. Fading distribution tends to be of Rayleigh's.

Microcells are widely employed in UMTS for full coverage, mainly in urban areas. Compact BSs are usually mounted on lampposts and buildings in the neighborhood. Microcells have an RMS delay spread ranging from 10 to 100 ns, which allows maximum transmission rate of 1 Mbits/s. The distribution that characterizes fading tends to be of Nakagami-Rice's (Andersen et al., 1995).

Picocells are used to provide services for areas with high terminal density as well as for in-building mobile communication systems.

Umbrella cells are used to maintain coverage continuous and to assist in the handover of terminals moving in picocells at full speed.

Finally, road cells are used to provide coverage for road sections that use compact BSs. The antennae used have their transmission pattern designed for the environment of highways (Cheung et al., 1994).

3.6.6 Detailed Planning for the Final Cost Estimation

At this stage, the BS location for each cell is determined and parameters such as antenna height, antenna gain and transmitter power are adjusted based on the particular characteristics of the environment, such as the irregularities of the terrain and coverage area in each cell. The results from the stages before satisfy coverage performance and try to meet traffic requirements.

However, there may still exist traffic overload in some cells. In order to obtain a reduced number of channels, it is possible to use channel allocation strategies that may be applied to cell system network planning so as to supply suitable capacities for the expected traffic, maintaining radio interference to a minimum.

If the expected system performance is reached after the adjustment, the final system design is obtained, and overall cost can be estimated. On the other hand, the results at this stage are fed into the previous stages, and the process is repeated. If that happens, the number of cells should be raised in order to satisfy the expected quality of service.

4

Base Station Deployment
based on Artificial Immune Systems

4.1 Introduction

The base station (BS) deployment problem is a multi-objective task. The best sites among a set of candidate sites are selected based on traffic hold requirements and propagation information for a given area. For each selected site, the number of installed antennae as well as their configurations (type of antenna, azimuth, tilt, gain, power, etc.) are determined. It involves many conflicting objectives: allow handoff between cells and at the same time guarantee minimum interference, produce the best coverage at the lowest possible cost of deployment and maintenance, respect the local environmental legislation and obey the rules imposed by the local government or by service regulatory agencies.

The simulation models already proposed try to balance computational speed and level of detail. There is always a trade-off between computational speed and level of detail. Increasing the realism of a model produces time-consuming solutions. The models show differences in terms of detail and complexity, in the simulation environment, in the type of network planning adopted, in the number of objectives and which objectives are considered, in the propagation model used for the path-loss calculation and in the optimization algorithm.

Many optimization algorithms have already been proposed: simulated annealing (Hao et al., 1997); local search algorithm (Rawnsley and Hurley, 2000); evolutionary algorithm (Allen et al., 2001); genetic algorithm (Huang et al., 2000), (Raisanen, 2007); tabu search (Vasquez and Hao, 2001).

The multi-objective optimization algorithms based on artificial immune systems (MO-AISs) are a new class of evolutionary algorithms. They are inspired by processes that take place in the human immune system, such as: affinity maturation, antigen recognition, and receptor editing (Castro and Timmis, 2002).

Many multi-objective algorithms based on immunological mechanisms have been proposed (Campelo et al., 2007; Coello et al., 2007): constrained multi-objective immune algorithm (CMOIA); multi-objective immune system algorithm (MISA; Cortes and Coello, 2005); vector immune system (VIS; Freschi, 2006); multi-objective clonal selection algorithm (MOCSA; Guimarães et al., 2007). In all MO-AIS algorithms, local, and global searches are carried out simultaneously (Luh and Chueh, 2004).

4.2 Multi-Objective Optimization

The multi-objective optimization problem or multiple criteria optimization problem is described as follows.

Find

$$\mathbf{x} = [x_1, x_2, \ldots, x_n]^T, \quad \mathbf{x} \in \Omega, \tag{4.1}$$

that minimizes or maximizes the objective function

$$f(\mathbf{x}) = [f_1(\mathbf{x}), f_2(\mathbf{x}), \ldots, f_x(\mathbf{x})]^T, \tag{4.2}$$

subject to

$$g_i(\mathbf{x}) \leq 0 \text{ para} \quad i = 1, \ldots, m, \tag{4.3}$$

and

$$h_j(\mathbf{x}) = 0 \text{ para} \quad j = 1, \ldots, p, \tag{4.4}$$

in which $\mathbf{x}$ is the vector of decision variables, m is the number of inequality constraints, p is the number of equality constraints and Ω contains all possible $\mathbf{x}$ that can be used to evaluate $f(\mathbf{x})$ (decision space). Equations (4.3) and (4.4) describe the dependencies among decision variables and parameters in the problem.

The goal of a multi-objective optimization problem is to find a solution that balances the conflicting objectives and constraints. This solution is not unique and cannot be considered a global optimum. The set of plausible solutions (or candidate solutions) is called the Pareto optimal set and form the Pareto front. Once the Pareto front is known, a decision maker is able to choose the most suitable solution for a given problem. The formal definitions of Pareto optimal set and Pareto front are provided next.

(Pareto Dominance) A vector $\mathbf{u} = [u_1, u_2, \ldots, u_k]^T$ dominates a vector $\mathbf{v} = [v_1, v_2, \ldots, v_k]^T$ $(\mathbf{u} \preceq \mathbf{v})$ if and only if $\mathbf{u}$ is partially less than $\mathbf{v}$, i.e., $\forall i \in \{1, \ldots, k\}, u_i \leq v_i \wedge \exists i \in \{1, \ldots, k\} : u_i < v_i$.

(Pareto optimality) A vector $\mathbf{x}^* \in \Omega$ is a Pareto optimal if and only if there isn't a vector $\mathbf{x}' \in \Omega$ for which $\mathbf{v} = f(\mathbf{x}') = [f_1(\mathbf{x}'), f_2(\mathbf{x}'), \ldots, f_k(\mathbf{x}')]^T$ dominates $\mathbf{u} = f(\mathbf{x}^*) = [f_1(\mathbf{x}^*), f_2(\mathbf{x}^*), \ldots, f_k(\mathbf{x}^*)]^T$. A non-dominant solution is Pareto optimal.

(Pareto optimal set) For a given multi-objective problem $f(\mathbf{x}) = [f_1(\mathbf{x}), f_2(\mathbf{x}), \ldots, f_k(\mathbf{x})]^T$, the Pareto optimal set P^* is defined as

$$P^* := \{\mathbf{x} \in \Omega | \neg \exists \mathbf{x}' \in \Omega f(\mathbf{x}') \preceq f(\mathbf{x})\}. \tag{4.5}$$

When a vector $\mathbf{x}$ that is a Pareto optimal is evaluated by $f(\mathbf{x})$, the vector $\mathbf{u}$ é obtained. The components of the vector $\mathbf{u}$ are the optimal solutions for each of the optimizing objectives. The performance of each component cannot be improved without affecting another. The set of all solution vectors that are Pareto optimals form the Pareto optimal set P^*.

(Pareto front) For a given multi-objective problem $f(\mathbf{x}) = [f_1(\mathbf{x}), f_2(\mathbf{x}), \ldots, f_k(\mathbf{x})]^T$ with the Pareto optimal set P^*, the Pareto front PF^* is defined as

$$PF^* := \{\mathbf{u} = f(\mathbf{x}) | \mathbf{x} \in P^*\}. \tag{4.6}$$

The Pareto front PF^* comprises the vectors obtained when the solutions that form the Pareto optimal set P^* are evaluated. All vectors in the Pareto front are non-dominant. As it is sometimes difficult to obtain the Pareto front when real engineering problems are considered, approximations are used. It is necessary to make a distinction between the real Pareto front PF_r^* and the approximated Pareto front PF_a^*, obtained through the optimization procedure.

4.2.1 The Multi-Objective Optimization Algorithms based on Artificial Immune Systems

Most MO-AISs are inspired by the clonal selection theory (Luh and Chueh, 2004; Cortes and Coello, 2005; Guimarães et al., 2007). The immune cells B go through a process called clonal expansion. The clonal expansion includes adaptation through mutation (somatic hyper-mutation) and a selection mechanism. This selection mechanism makes sure that the B cells, which produce antibodies with more affinity, survive, and subsequently become memory cells. This combination of mutation and selection is called the affinity maturation of the immune system.

Other concepts or theories usually applied to the development of artificial immune systems for multi-objective optimization are (Guimarães et al., 2007): the immune network theory, receptor editing through a DNA library and lymphokine or chemical messages.

The MO-AIS algorithms include a memory population or offline population and is divided into the following phases: affinity evaluation, avidity evaluation, selection for cloning, proliferation and mutation (somatic hypermutation) and diversification (Campelo et al., 2007). The iterative process is repeated several times until a stopping criterion is met. The offline population is constantly updated in the process.

The memory population or offline population stores the best solutions, the Pareto front approximation. Dominance relations are used to compare the vectors obtained throughout the iterative process.

In optimization, affinity means the evaluation of the objective function $f(\mathbf{x})$ and the constraints (4.3) and (4.4). Avidity refers to the overall binding intensity between an antigen $f(\mathbf{x})$ and an antibody (the solution vector $\mathbf{x}$). Therefore, it measures the quality of the candidate solution. In MOCSA, the candidate solutions (antibodies) are classified into successive non-dominant fronts according to dominance relations.

The selection for cloning the best N_c might be performed deterministically or stochastically. In order to promote selection proportional to the affinity between antibodies and antigens or selection according to avidity, any selection mechanism commonly used in evolutionary algorithms might be used, such as: roulette wheel selection, elitist selection, hierarchical selection, tournament selection and bi-classist selection.

The mutation rate inversely proportional to affinity and the number of clones proportional to affinity when combined result in a balance of local and global searches. Once the percentage of individuals to be cloned is chosen, the number of clones that each produces may be determined in different ways. The same number of clones for each antibody is usually applied (Castro and Timmis, 2002). The greater the number of individuals proliferating, the longer is the processing time. In MOCSA, the number of clones is determined based on the Pareto front the antibodies belong (Guimarães et al., 2007). The number of clones is given by

$$N_c = \text{round}\left(\frac{\beta N}{i}\right),\tag{4.7}$$

in which β is a multiplying constant, N is the overall number of antibodies, round $(\cdot)$ is an operator that returns the closest integer value to its argument and i is the number of the Pareto front it belongs.

The diversification phase is related to the global searches. It is not present in all MO-AIS algorithms. By applying diversification, it is possible to add new solutions randomly. Based on the immune network theory, a common suppressing operator is usually applied (Freschi, 2006). When two antibodies are too close to each other, one of them might recognize the other and, therefore, one of them is eliminated. If the Euclidean distance between two antibodies in the objective space is greater than a given value ε_1, the antibody with the greatest affinity will be suppressed.

In MOCSA, suppression is applied both to the decision variable space and to the objective space (Coello et al., 2007). In a more recent work, Campelo et al. only applied suppression to the objective space (Guimarães et al., 2007). The objective space vectors are first normalized to the unitary hypercube to account for possible discrepancies between threshold values for each objective. Then the distances between the remaining antibodies N in the offline population are calculated. The distances of each individual to its k closest neighbors are obtained, in which k is given by

$$k = \text{round}\left(\sqrt{N}\right). \tag{4.8}$$

The individual with the smallest sum of the k distances is eliminated, as it is located in a dense region of the Pareto front. The procedure is repeated until the memory population reaches the maximum size specified by the user.

4.3 Proposed Formulation

The proposed model is a discrete test point model based on (Raisanen, 2007). Test point models have several advantages over other approaches. They simplify the measurement of all network objectives, the candidate sites may be freely chosen and the models are easily adaptable both to greenfield and to network expansion scenarios. The working area or simulation area W is discretized into test points at Cartesian coordinates (x, y, z) at a given resolution. The following data is defined:

- The reception test points, **RTPs**, where signal reception quality is measured;
- The service test points, **STPs**, where the received signal must be above the service threshold S_q, to ensure the quality of service is met;
- The traffic test points, **TTPs**, which each carrying a traffic load in erlang;
- The candidate sites, **CBS**, which may contain up to three antennae;
- The angle of incidence matrix, **AIM**, that specifies the vertical angles from each **CBS** to each **RTP**;
- The path-loss matrix, **PLM**, with information regarding the signal loss from each **CBS** to each **RTP**.

The standard urban empirical model proposed by Hata is used for path-loss calculation. The model also considers random shadowing effects as proposed by Huang et al. (Huang et al., 2000). These can either amplify or attenuate the strength of the signal at reception. They are obtained by the next pseudo-random Gaussian value (μ, σ), in which μ is the path-loss value and σ is 4 dB.

The best server model is adopted, in which each **STP** is served by the CBS providing the greatest received signal strength. A cell is defined by the set of STPs covered by one antenna, in which $P_r \geq -90$ dBm. Each site may contain up to three antennae.

A network subset **CBS'** refers to a set of sites **CBS** with at least one active antenna and satisfies the following objectives:

- Coverage – It is the sum of the STP_i covered in the working area divided by the total number of **STP** as a percentage. Thus,

$$\text{COVER}_{\text{CBS}'} = \frac{\sum_{i=1}^{n_{\text{STP}}} STP_i}{n_{\text{STP}}} \times 100, \tag{4.9}$$

in which

$$STP_i = \begin{cases} 1, & \text{if } STP_i \text{ is covered,} \\ 0, & \text{otherwise;} \end{cases} \tag{4.10}$$

- Unit cost – It is the number of sites with at least one active antenna. Thus,

$$\text{COST}_{\text{CBS}'} = \sum_{\text{CBS}_i \in \text{CBS}'} CBS_i; \tag{4.11}$$

- Traffic hold – It is the sum of the current overall traffic in the network divided by the total traffic load and expressed as a percentage. Thus,

$$\text{TRAF}_{\text{CBS}'} = \frac{\sum_{i=1}^{n_{\text{CBS}'}} T_{\text{CBS}'_i}}{\sum_{i=1}^{n_{\text{TTP}}} TTP_i} \times 100. \tag{4.12}$$

The optimization strategy is divided into three phases: pre-processing, site initialization, and iterative process. In the pre-processing phase, the configuration of the antennae that ensure that maximum traffic load in each cell is no >5.4 erlang and the set of STPs covered by each site are determined. The key aspect of the approach is to perform cell dimensioning only once and optimize the network for high service coverage and low cost. If all STPs are covered, all TTPs are covered by definition, since $TTP \subseteq STP \subseteq RTP$.

Site initialization with binary representation speeds up the iterative process and reduces computational time (Raisanen, 2007). Each individual in the population is identified by a binary string with length n_{CBS} (number of candidate sites). Whether a site is on or off is indicated, respectively, by 1 or 0. The initial population is randomly generated and the number of active sites is chosen according to the following criterion: For the first third of the population, the number of active sites is chosen between 1 and the minimum number of sites, which could satisfy 100% traffic hold; For the second third of the population, the total number of sites to be turned on is between the minimum and double the minimum.

The optimizing algorithm is based on MOCSA (Guimarães et al., 2007). MOCSA adopts real-valued variables and is inspired by the clonal selection theory and the immune network theory. MOCSA produced satisfactory results when compared to other MO-AIS algorithms. The binary-coded multi-objective optimization algorithm (BRMOA) uses binary representation for the decision variables and replaces the Gaussian mutation, adopted by MOCSA, with a uniform mutation (Castro and Timmis, 2002), with probability

$$p_m = \frac{1}{\sqrt{n_{CBS}}}. \tag{4.13}$$

4.4 Simulation Results

The simulation environment mimics a metropolis where traffic is randomly distributed and was proposed by Raisanen (2007). Figure 4.1 shows the simulation environment and Table 4.1 contains the simulation environment data.

Binary-coded multi-objective optimization algorithm has the following user-defined parameters: the size of the initial population P_0, the number of generations g_{max}, the maximum number of evaluations of the objective function n_e, the size of the memory population P_m and the percentage of individuals replaced in every generation d.

In large-scale optimization problems, such as the BS deployment problem, in which a large number of variables is considered, computational speed as well as the quality of the solutions play an important role. BRMOA keeps a record of each individual already evaluated. Therefore, it is possible to reduce the number of objective function evaluations. In other to evaluate the quality of the Pareto front approximations, many aspects should be considered: the closeness to the real Pareto front, the uniform spread and the diversity of vectors along the Pareto front. Many metrics or quality indicators have already been proposed (Coello et al., 2007).

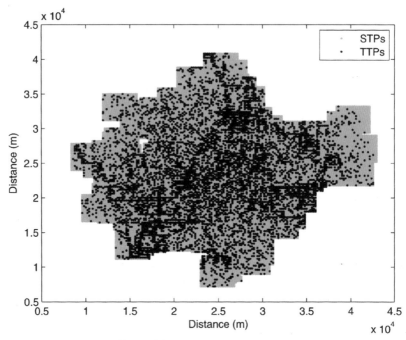

Figure 4.1 Simulation environment.

Table 4.1 Simulation environment

Working Area	2,252.64 km^2
Reception test points (RTPs)	56,792
Service test points (STPs)	17,393
Candidate sites	568
TTPs	6,602
Traffic load	3,221.84 erlang

Schott's spacing metric was chosen as a quality indicator (Coello et al., 2007). It does not require the researcher to know the real Pareto front. It measures the distance variance of neighboring vectors in the Pareto front and is given by,

$$S_{\mathrm{p}} = \sqrt{\frac{1}{N-1} \sum_{i=1}^{N} \left(\overline{d} - d_i \right)^2}, \qquad (4.14)$$

$$d_i = \min_j \left| f_1^i \left(\overrightarrow{x} \right) - f_1^j \left(\overrightarrow{x} \right) \right| + \left| f_2^i \left(\overrightarrow{x} \right) - f_2^j \left(\overrightarrow{x} \right) \right|, \qquad (4.15)$$

in which N is the number of vectors in PF*, $\overline{d}$ is the mean of all d_i and $f_k^i(\overrightarrow{x})$ is each component of the objective function.

When the spacing is null, the algorithm has found the ideal distribution of non-dominated vectors, i.e., all vectors are uniformly distributed.

The Pareto front approximations were obtained for initial populations of 10, 15, and 30 individuals, when: $g_{max} \leq 10$, $n_e = 12.000$, $P_m = 100$ **e** $d = 0.25$. As BRMOA is a stochastic optimizer, a statistical analysis was carried out based on 10 runs of the software. Computing time was estimated based on the number of objective function evaluations. The parameters P_0 and g_{max} were chosen according to the number of objective function evaluations and the required number of solutions in the Pareto front.

Table 4.2 shows the spread of the vectors for initial populations of 10, 15, and 30 individuals after 10 generations and in 10 runs of the software. The results show that better Pareto front approximations are obtained when more generations are considered. The best approximations were only obtained after 10 generations. As the global searches are intensified, subtle differences are observed in the spacing metric. This is shown in the standard deviation. According to the spacing metric, a satisfactory approximation is obtained with an initial population of 10 individuals and after 10 generations.

Figure 4.2 shows the number of objective function evaluations after each generation and Figure 4.3 shows the number of vectors in the Pareto front after each generation when initial populations of 10, 15, and 30 individuals are considered. The number of objective function evaluations is always high when a population of 30 individuals is considered, even in the first generations. An initial population of 30 individuals is recommended when there is little concern about computing time. On the other hand, a initial population of 30 individuals ensures more vectors in the Pareto front after 10 generations and 2,989 objective function evaluations. After 10 generations the same number of vectors in the Pareto front is obtained when initial populations of 10 or 15 individuals are considered. From the sixth generation onward, it is possible to obtain 60 vectors in the Pareto front and after 1,810 objective

Table 4.2 Spacing for initial populations of 10, 15, and 30 individuals after 10 generations and in 10 runs of the software

Spacing	10 Individuals	15 Individuals	30 Individuals
Mean	2.7272	2.2032	2.1828
Minimum	2.2015	2.0349	1.7329
Maximum	3.2528	2.3067	2.8453
Standard deviation	0.7434	0.1470	0.1164

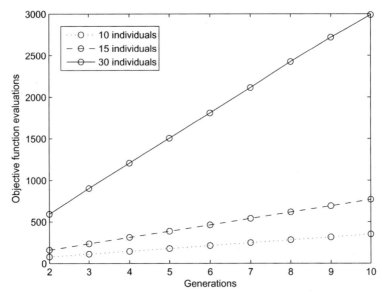

Figure 4.2 Number of objective function evaluations after each generation for initial populations of 10, 15, and 30 individuals.

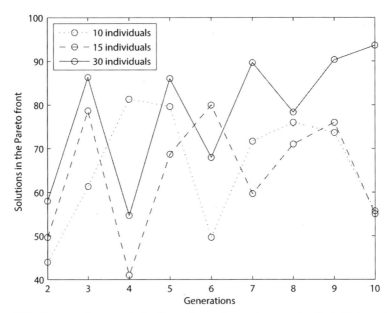

Figure 4.3 Number of vectors in the Pareto front after each generation for initial populations of 10, 15, and 30 individuals.

function evaluations. It is important to point out that the number of solutions is sometimes reduced sharply in two or more successive generations. This is most noticeable when initial populations of 10 and 15 individuals are considered. This is due to the suppression operator that eliminates similar or very close individuals from the memory population. Even though local optima still appear when a population of 30 individuals is considered, the oscillations are less frequent. These oscillations are related to the diversity mechanism in the MO-AIS algorithms, when new global searches are performed.

Figure 4.4 shows the best Pareto front approximations for initial populations of 10, 15, and 30 individuals. The costs of deployment for total coverage of the working area are very high. The costs are reasonable when a coverage between 90 and 95% of the working area is considered. An initial population of 15 individuals produces satisfactory results after 463 objective function evaluations. The best approximation is obtained when an initial population of 30 individuals is adopted and after 10 generations (approximately 3,000 objective function evaluations).

Table 4.3 shows coverage, unit cost, traffic hold and number of objective function evaluations for the solution that produces maximum coverage when different initial populations are considered and after 10 runs of the software. Maximum coverage is obtained when an initial population

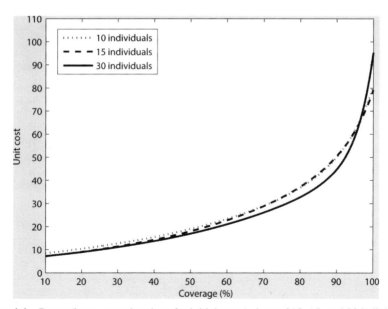

Figure 4.4   Pareto front approximations for initial populations of 10, 15, and 30 individuals.

Table 4.3   Coverage, unit cost, traffic hold and number of objective function evaluations for the solution that produces maximum coverage

Initial Population	Coverage		Cost		Traffic		Evaluations	
	Mean	Standard deviation (std. dev.)	Mean	Std. dev.	Mean	Std. dev.	Mean	Std. dev.
10 Individuals	96.470	0.707	68.670	6.028	96.200	1.304	284.000	110.869
15 Individuals	97.376	0.284	69.667	9.504	97.64	0.512	312.000	146.010
30 Individuals	99.396	0.179	93.000	4.000	99.360	0.246	2,408.000	835.900

of 30 individuals is adopted, after 8 generations. By choosing an initial population of 30 individuals, coverage is increased by 3.03%. Maximum coverage also implies higher deployment costs. The value obtained for maximum coverage is greater and more accurate than the value obtained in (Raisanen, 2006) by applying the genetic algorithm NSGA-II. Less computing time is also obtained, since Raisanen considered an initial population of 1,200 individuals and 50 generations.

Figures 4.5, 4.6, 4.7, and 4.8 show sample solutions at 95% coverage, 98% coverage, maximum coverage, and less deployment cost with reasonable coverage. The decision maker is able to choose the most suitable solution.

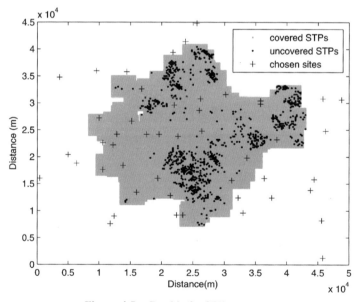

Figure 4.5   Graphic for 95% coverage.

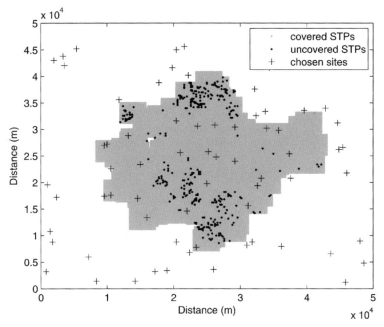

Figure 4.6 Graphic for 98% coverage.

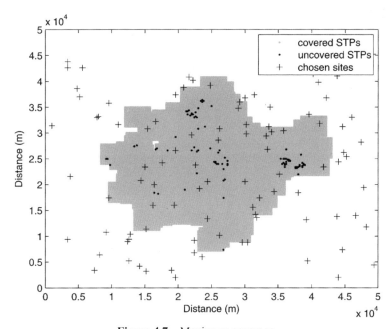

Figure 4.7 Maximum coverage.

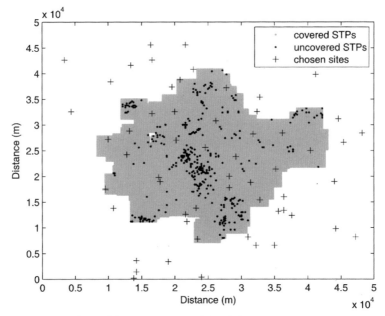

Figure 4.8   Lowest cost with satisfactory coverage.

When technical requirements are the main concern, the solutions at higher coverage are to be preferred. The solutions at 95% coverage and low cost show the problematic areas in terms of coverage, either because of great traffic load or because of the topography of the terrain that hinders propagation.

5

Cost-Effective Base Station Deployment Approach based on Artificial Immune Systems

Some minor changes are made to the binary-coded multi-objective optimization algorithm (BRMOA), in order to obtain the cost-effective base station (BS) deployment approach. Instead of considering unit cost as described previously, cost is defined as follows:

$$\text{COST}_{\text{CBS}'} = \sum_{\text{CBS}_i \in \text{CBS}'} [C_{\text{f}}(CBS_i) + C_{\text{g}}(CBS_i)], \qquad (5.1)$$

in which $C_{\text{f}}(CBS_i)$ is the fixed cost of deploying a BS at a plain site and $C_{\text{g}}(CBS_i)$ is the geographical cost.

The fixed cost of deploying a BS at a plain site includes:

- Acquisition, shipping, and installation of equipment;
- Software licenses;
- Site acquisition or rent in an area, in which there are no restrictions on BS deployment, regarding environmental impact, radio emission, or local environmental legislation;
- Site legalization (bureaucratic fees, administrative fees for permission for radio emission, legal expenses, technical reports from experts in the field in order to approve the site location);
- Site preparation including construction.

The geographical cost depends on the site location. It may slightly increase overall costs or rather hinder site selection. The working in area is divided into the following categories:

- Standard or plain area:
  It comprises urban and suburban areas where there are no restrictions on radio emission;

- Areas with surcharge:
  The surcharge refers to the high cost of property in the city centre or in any other high-priced residential areas and also to higher deployment costs in rural areas;
- Prohibited areas:
  Even though environmental legislation tends to limit BS deployment in certain areas, the companies may still need to choose a prohibited site so as to meet technical requirements or for business expansion purposes. Choosing a prohibited site means extra cost related to fines and other bureaucratic fees;
- Preferred areas:
  In order to reduce deployment costs, preferred areas are chosen. These include areas where site sharing is possible and buildings or areas belonging to the company or business partners. The local government might as well lease areas or buildings for deployment at reduced costs;
- Mandatory areas:
  These include areas where site deployment is mandatory for security reasons (inside tunnels and under overpasses) and areas where a large number of people are usually gathered (football stadiums, close to theatres, shopping malls, convention, and trade centres, etc.);
- Non-regulated areas:
  These are not included in any of the previous categories and new procedures or fees may apply. In non-regulated areas, the geographical cost is a random value between the fixed cost of deploying a BS at a plain site and the geographical cost of deploying BSs in prohibited areas.

The optimization strategy is divided into three phases: pre-processing, site initialization, and iterative process. In the pre-processing phase, the configuration of the antennae that ensure that maximum traffic load in each cell is no greater than 5.4 erlang and the set of service test points (STPs) covered by each site are determined. The key aspect of the approach is to perform cell dimensioning only once and optimize the network for high service coverage and low cost. If all STPs are covered, all traffic test points (TTPs) are covered by definition, since $TTP \subseteq STP \subseteq RTP$.

Site initialization with binary representation speeds up the iterative process and reduces computational time (Raisanen, 2007). Each individual in the population is identified by a binary string with length n_{CBS} (number of candidate sites). Whether a site is on or off is indicated, respectively, by 1 or 0.

The initial population is randomly generated and the number of active sites is chosen according to the following criteria:

- For the first third of the population, the number of active sites is chosen between 1 and the minimum number of sites, which could satisfy 100% traffic hold.
- For the second third of the population, the total number of sites to be turned on is between the minimum and double the minimum.

The optimizing algorithm is based on multi-objective clonal selection algorithm (MOCSA; Guimarães et al., 2007).

MOCSA adopts real-valued variables and is inspired by the clonal selection theory and the immune network theory.

MOCSA produced satisfactory results when compared to other multi-objective optimization algorithms based on artificial immune systems (MO-AISs) algorithms. The BRMOA uses binary representation for the decision variables and replaces the Gaussian mutation, adopted by MOCSA, with a uniform mutation (Castro and Timmis, 2002), with probability

$$p_{\mathrm{m}} = \frac{1}{\sqrt{n_{\mathrm{CBS}}}}. \tag{5.2}$$

The proposed model is very flexible when it comes to candidate sites. It is possible to choose which sites to activate throughout the iterative process. The user may use his experience to interfere in the process and lead the search to previously selected sites.

5.1 Simulation Results

The simulation environment used for analysis mimics a metropolis. Figure 5.1 shows the simulation Table 5.1 contains the simulation environment data. The candidate sites are divided into twelve categories according to different geographical costs as show in Table 5.2. The standard or plain area accounts for 58.68% of the working area with randomly distributed candidate sites. Some preserved and high-priced residential areas are expected to handle high traffic load. Therefore, two prohibited sites are needed to meet traffic requirements. Preferred sites are randomly distributed in the working area. Both non-regulated and rural areas are far away from the city centre. Geographical costs are userdefined.

Two scenarios are considered for the analysis. In the first scenario, sites in mandatory areas are active during the whole iterative process. On the other hand, the sites in preserved areas are always disabled. The remaining

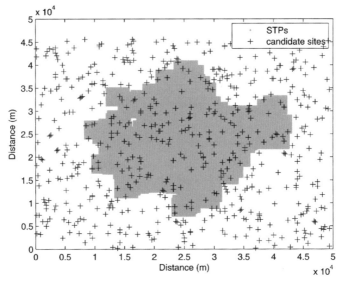

Figure 5.1   Simulation environment.

Table 5.1   Simulation environment data

Working Area	2,252.64 km^2
Reception test points (RTPs)	56,792
STPs	17,393
Candidate sites	568
TTPs	6,602
Traffic load	3,221.84 erlang

sites are activated or disabled according to simulation results. In the second scenario, sites in mandatory areas as well as sites in preferred areas are always active. The analysis is carried out in order to find out if total coverage is obtained, which sites are selected and the criteria used to choose sites in both scenarios.

Table 5.3 describes the two scenarios. Minimum cost and minimum number of selected sites refer to active sites before the optimization strategy is applied.

The Pareto front approximations were obtained for an initial population of 30 individuals, when: $g_{max} \leq 10$, $n_e = 12,000$, and $P_m = 100$ e $d = 0.25$. As BRMOA is a stochastic optimizer, a statistical analysis was carried out based on ten runs of the software. Computing time was estimated based on the number of objective function evaluations. Schott's spacing metric was chosen as a quality indicator (Coello et al., 2007). It does not require

Table 5.2 Number of candidate sites and geographical cost according to the different categories

	Candidate Sites	C'_g (per site)
Standard areas		
Urban and suburban areas	339	0
Areas with surcharge		
City centre	7	1.00
High-priced residential areas	37	2.00
Rural areas	31	1.00
Optional areas		
Non-regulated areas	67	1.00–9.00
Prohibited areas		
Preserved areas	44	9.00
Preferred areas		
Site sharing	20	−0.50
Business partners	4	−0.50
Government	3	−0.25
Mandatory areas		
Indoor (shopping malls)	7	4.00
Indoor (tunnels and overpasses)	3	3.00
Crowded areas	6	1.00

Table 5.3 Description of both scenarios

	Scenario 1	Scenario 2
Preserved areas	2 Active sites	2 active sites
Preferred areas	"on" or "off"	Always "on"
Mandatory areas	Always "on"	Always "on"
Sites in other areas	"on" or "off"	"on" or "off"
Minimum sites	18	45
Minimum cost	77.00	91.25

the researcher to know the real Pareto front. It measures the distance variance of neighbouring vectors in the Pareto front and is given by,

$$S_{\mathrm{p}} = \sqrt{\frac{1}{N-1} \sum_{i=1}^{N} \left(\overline{d} - d_i\right)^2}, \tag{5.3}$$

$$d_i = \min_j \left| f_1^i\left(\overrightarrow{x}\right) - f_1^j\left(\overrightarrow{x}\right)\right| + \left| f_2^i\left(\overrightarrow{x}\right) - f_2^j\left(\overrightarrow{x}\right)\right|, \tag{5.4}$$

in which N is the number of vectors in PF*, $\overline{d}$ is the mean of all d_i and $f_k^i(\overrightarrow{x})$ is each component of the objective function.

Figure 5.2 shows the best Pareto front approximations according to Schott's spacing metric for an initial population of 30 individuals for the first scenario.

Figure 5.3 shows the Pareto front approximations for the second scenario. The minimum network configuration guarantees coverage at 76.44% in the first scenario and coverage at 96.24% in the second scenario. Total coverage is not obtained in any of the scenarios. In the first scenario, the maximum coverage is 99.72%, which refers to deployment cost of 207.00. In the second scenario, the maximum coverage is 99.97%, with deployment cost of 204.00. There is an increase of 0.26% at coverage and overall deployment cost is reduced by 1.45% when the second approach is adopted. The addition of preferred sites with low deployment cost guarantees satisfactory coverage even before the optimization procedure is carried out. Overall deployment cost is higher for the first scenario, when the same coverage is attempted.

Figure 5.4 shows the number of objective function evaluations after each generation for both scenarios. More objective function evaluations are observed in the first scenario, since more variables are considered. Figure 5.5 shows the number of vectors in the Pareto front after each generation for

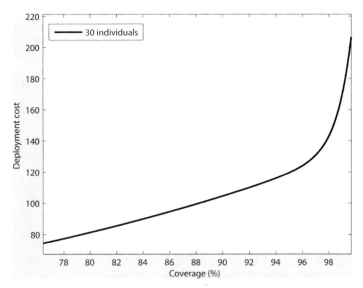

Figure 5.2   Pareto front approximation for scenario 1.

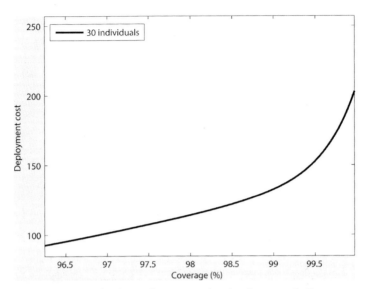

Figure 5.3   Pareto front approximation for scenario 2.

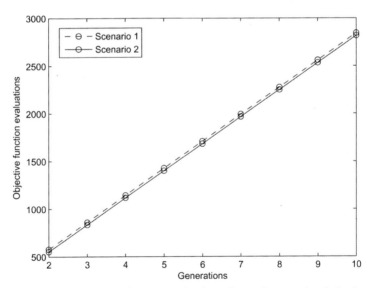

Figure 5.4   Number of objective function evaluations after each generation for both scenarios.

both scenarios. The number of solutions is sometimes reduced sharply in two successive generations. This is related to the suppression operator that eliminates similar or very close individuals from the memory population.

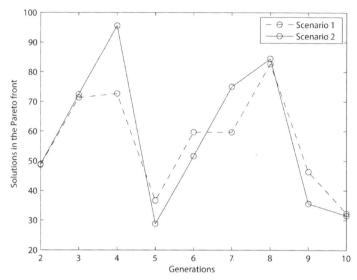

Figure 5.5   Number of solutions in the Pareto front after each generation for both scenarios.

The oscillations in the number of individuals in the memory population are more intense in the second scenario, although the number of individuals is the same after ten generations. It is important to point out that the number of solutions in the Pareto front is higher than 30 after the very first generation for both scenarios.

Tables 5.4 and 5.5 show coverage, cost, number of sites, and traffic for solutions at 97, 98, and 99% coverage for both scenarios. The results were obtained in 10 runs of the software and after only one generation. When the number of candidate sites is restricted, maximum coverage is attained faster as shown in Table 5.4.

Table 5.4   Coverage, cost, number of sites and traffic at 97, 98 e 99% coverage for scenario 1

Coverage	Cost	Sites	Traffic
97.026	141.167	60.667	96.717
0.204	6.449	1.528	0.639
97.984	152.583	67.667	98.101
0.098	8.064	3.055	0.324
98.948	181.167	74.000	99.064
0.168	20.642	10.001	0.295

Table 5.5 Coverage, cost, number of sites and traffic at 97%, 98% e 99% coverage for scenario 2

	Coverage	Cost	Sites	Traffic
Mean	96.901	101.583	51.333	96.424
Standard deviation	0.304	1.528	2.517	0.443
Mean	98.244	119.25	64.000	98.101
Standard deviation	0.295	3.606	2.000	0.581
Mean	99.047	138.917	69.667	99.185
Standard deviation	0.009	5.033	1.528	0.131

Tables 5.6 and 5.7 show the type of candidate sites selected for solutions at 97% coverage, 98% coverage, 99% coverage, and maximum coverage. In the first scenario, low-cost preferred areas are only selected in order to

Table 5.6 Selected sites in different solutions for scenario 1

	Coverage			
Types of Candidate Sites	97%	98%	99%	Max.
Standard urban areas	31	41	41	43
City centre	2	1	1	2
High-priced urban areas	3	4	3	6
Rural areas	1	2	3	7
Non-regulated areas	2	2	6	4
Preserved areas	2	2	2	2
Preferred areas	2	3	2	4
Mandatory areas	16	16	16	16
Total	59	71	74	84

Table 5.7 Selected sites in different solutions for scenario 2

	Coverage			
Types of Candidate Sites	97%	98%	99%	Max.
Standard urban areas	5	18	19	42
City centre	0	0	1	0
High-priced urban areas	0	1	1	9
Rural areas	0	0	3	3
Non-regulated areas	1	1	1	8
Preserved areas	2	2	2	2
Preferred areas	27	27	27	27
Mandatory areas	16	16	16	16
Total	51	66	70	107

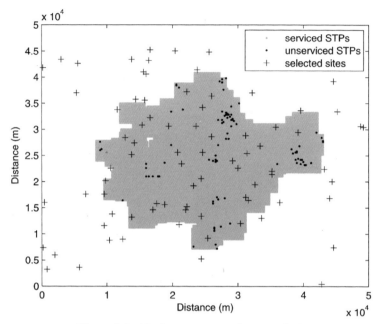

Figure 5.6   Maximum coverage for scenario 1.

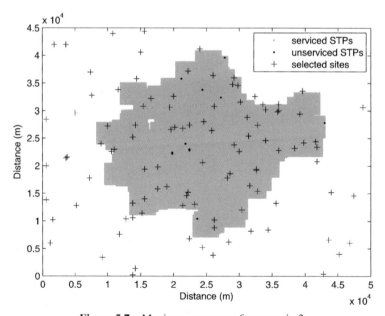

Figure 5.7   Maximum coverage for scenario 2.

reduce deployment costs. When higher coverage is attempted, standard sites are chosen. The optimization strategy selects areas with surcharge based on technical requirements. Non-regulated areas are only selected when their geographical costs are low. In the second scenario, there is a sharp fall in the number of sites in standard areas. High-cost sites are only activated so as to meet traffic requirements. Figures 5.6 and 5.7 show sample solutions at maximum coverage for both scenarios.

6

Cell Planning Using Voronoi Diagrams

Network planning is one of the foundations of a cellular project, and can be done by the operator, or by the vendor, as a paid service. Generally, it involves the planning of the radio network, of the transmission and of the core network. Some of the selection criteria for site selection and acquisition are the structural engineering adequacy, the acquisition feasibility, the availability of line-of-sight, the subscriber potential, legal or geometrical restrictions, and radiofrequency planning. (Mishra, 2007).

The mobile cellular network is normally represented by means of hexagonal topology. This structure is useful for planning frequency reuse but not appropriate for the analysis of handoff and coverage. This chapter presents the cellular network as a dynamic, considering an ordered multiplicatively weighted Voronoi diagram (MWVD).

Radio parameters such as antenna height, transmission power, and specific-environment propagation characteristics are used as the basis to define the proximity rule in order to generate the Voronoi diagram. The cell boundaries are the edges of the Voronoi diagram. They are defined by comparison of the radii of adjacent cells. The proximity between a mobile and a base station (BS) is determined by means of an Euclidean distance weighted by propagation parameters.

6.1 Introduction

In a mobile cellular network, a Mobile Station (MS) is connected to the closest BS according to the system quality requirements. The connections are established taking into account the received power, the Signal to Noise Ratio (SNR) or the Bit Error Rate (BER).

Those parameters are determined by the transmit power, antenna heights, path loss and specific-environment propagation characteristics. The BS solves a closest-point problem (Shamos and Hoey, 1975) when connecting a mobile

133

to a BS. The BS radio signal power decreases with distance to a minimum level which defines the cell size.

This chapter presents the cellular coverage region by means of a Voronoi diagram, in which the cells are the partitions, called Voronoi regions, defined by a proximity rule (Basch et al., 1997). The cellular mobile system analyzes a number of probable MS–BS connections and chooses those with best quality. In a general way, the connection quality decreases with distance. Consequently, the connections are established on a "nearest neighbor" basis. The cell boundary is determined by a closest-point search and the mobiles lie in a partition.

6.2 The Voronoi Diagram

The Voronoi diagram is a geometric structure that assumes the proximity (nearest neighbor) rule in associating each point in the $\mathbb{R}^n$ space to a site point closest to it. This diagram is a partition set, generated by site points located inside each partition. Any point in the $\mathbb{R}^n$ space is associated to a site point by a proximity rule. Each partition is called a Voronoi region.

Let $\mathbf{x}$ be a point in the $\mathbb{R}^n$ space, $C = \{c_1, \ldots, c_N\}$ the site points set, $E(i,j)$ the edge between two site points and V_i the Voronoi region generated by $\mathbf{c}_i$. The proximity relation $x \in V_i$ occurs according to the proximity rule.

IF $\mathbf{x}$ is closer to $\mathbf{c}_i$ than any other site point THEN $\mathbf{x} \in V_i$.

Based on this proximity rule the Voronoi region is defined as

$$V_i = \{x | D(x, c_i) < D(x, c_j), \forall j \neq i\}, \tag{6.1}$$

in which the proximity metric D is a function of distance. According to the definition of D, several types of Voronoi diagram can be generated. The features of the Voronoi diagrams are extensively described in (Aurenhammer, 1991).

6.2.1 Types of Voronoi Diagrams

- **Generalized** – The proximity metric is $D = d$ in which d is the Euclidean distance. The Voronoi region is defined as

$$V_i = \{x | D(x, c_i) < D(x, c_j), \forall j \neq i\}. \tag{6.2}$$

  The edges between site points are straight lines.
- **Directional** – The proximity metric is a weighted distance whose weight is a function of the azimuth around the site point $D = d(\theta)$. The Voronoi region is defined as

$$V_i = \left\{ x \left| \frac{d(x, c_i)}{w_i(\theta)} < \frac{d(x, c_j)}{w_j(\alpha)}, \forall j \neq i \right. \right\}; \qquad (6.3)$$

- **Truncated** – This characteristic is expressed as a limit to the distance value

$$V_i = \left\{ x \left| \frac{d(x, c_i)}{w_i} < \frac{d(x, c_j)}{w_j}, \forall j \neq i, d \leq d_{\max} \right. \right\};$$

- **Dynamic** – This is a characteristic of the Voronoi diagrams. Inclusions and exclusions of site points can take place;
- **Multiplicatively weighted** – The proximity metric to the MWVD is $D = d/w$ where w is the weight of the corresponding site point. The Voronoi region is defined as

$$V_i = \left\{ x \left| \frac{d(x, c_i)}{w_i} < \frac{d(x, c_j)}{w_j}, \forall j \neq i \right. \right\} \qquad (6.4)$$

and the edges between site points are circular arcs. An example based on the following data is illustrated in Figure 6.1.

Site Point	c^1	c^2	c^3	c^4
Location	(−1,5)	(1,1)	(2.5,9)	(5,6)
Weight	3.6	4.0	3.2	2.8

- **Power diagram** – This is a generalization of the Voronoi diagram (Aurenhammer, 1987). The site points are defined as balls with center $\langle x_i, y_i \rangle$ and radius r_i. The proximity metric is $D = d^2 - r^2$ and the Voronoi region is defined as

$$V_i = \{ x | d^2(x, c_i) - r_i^2 < d^2(x, c_j) - r_j^2, \forall j \neq i \}.$$

The edges between site points are straight lines. However, in a pair of adjacent balls, the edge approaches the center of the ball with smallest radius. An example is shown in Figure 6.2.

6.3 The Order-k Voronoi Diagram

Let X be a point in $\mathbb{R}^n$, $C = \{c_1, \ldots, c_N\}$ the site points set, X a subset of C, $X \subseteq C$, and $\overline{X} = X - C$. The order of the Voronoi diagram is the cardinality of the subset X:$|X| = k$. An order-k Voronoi region is closer to

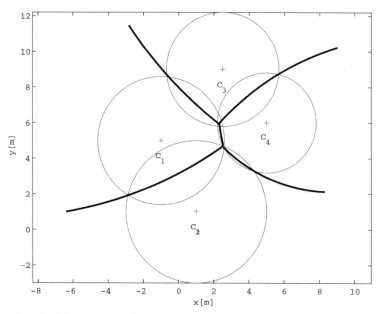

Figure 6.1   Multiplicatively weighted Voronoi diagram (MWVD) in the plane. The edges are circular arcs.

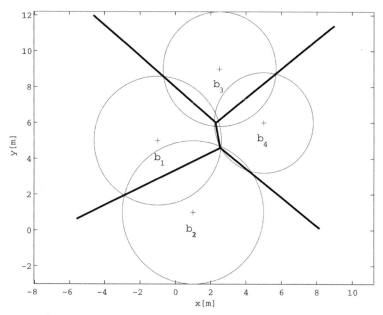

Figure 6.2   Power diagram in the plane. The edges are straight lines which approaches the ball with smallest radius.

the site points in X than any other site point in $\overline{X}$. The order-k Voronoi region $V(X)$ is defined as (Lee, 1982)

$$V(X) = \{x : D(x, c_i) < D(x, c_j), \forall c_i \in X, \forall c_j \in \overline{X}\},$$
$$\mathbb{R}^n = \bigcup_{X \subseteq C} V(X); k \in \mathbb{Z}^+; k < N.$$

The order-$(k + 1)$ Voronoi diagram is obtained from the extensions of the edges of the order-k diagram as follows:

- Remove a site point c_i obtaining new edges. The effect of removing c_i is the elimination of the edges determined by c_i and the extension of the neighbor edges to the interior of V_i;
- Place the previous removed site point c_i back into its location and remove another site point c_j;
- Repeat for all the site points.

Figure 6.3 presents an example of order -1, -2 and -3 generalized Voronoi diagrams.

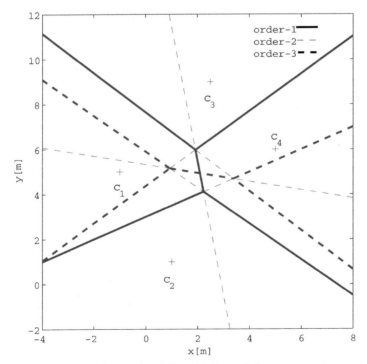

Figure 6.3 Generalized orders 1, 2, and 3 Voronoi diagrams superimposed.

6.3.1 The Ordered Order-k Voronoi Diagram

The ordered order-k MWVD is an order-k diagram in which the proximity relations are ordered in sequence of proximity. It is obtained from the superposition of the order-$k, k-1, \ldots, 1$ diagrams (Held and Williamson, 2004). The resulting spatial tessellation has the following features: (i) each Voronoi region represents the domain of a set of site points; (ii) proximity relations are ordered in sequence of proximity. A region denoted as $O(i, j, p)$ is the locus of all the points in $\mathbb{R}^n$ closest to the site point c_i and farthest to the site point c_p. The site points in O are ordered in a sequence of proximity.

6.3.2 Order-k Voronoi Diagrams Properties

Some properties of the Voronoi diagram defined to $\mathbb{R}^2$ are described as follows.

1. An edge is a bisector, it divides the plane into half-planes
$$E(i, j) = \pi_i \cap \pi_j;$$

2. The intersection of bisectors defines another bisector according to the operation
$$E(i, j) \oplus E(j, p) = E(i, p).$$
   The intersection $E(i, j) \cap E(j, p)$ generates the edge $E(i, p)$;

3. A vertex is defined as
$$\mathcal{V}_{ijp} = E(i, j) \cap E(j, p) \cap E(i, p);$$

4. The edges of the same order do not cross;

5. The maximum order of a diagram is limited to $k_{\max} = N_v + 1$, in which N_v is the number of vertices in the order-1 diagram. In the MWVD, the edges are circular arcs, thus vertices may not exist to disjoint regions, or a repetition of the same vertex may occur. One vertex should be taken on account only one time to obtain N_v which leads to the conclusion: $1 \le k_{\max} \le N_v + 1$;

6. The order-$(k + 1)$ diagram is obtained from that of order k. The edges expand in superior orders until cross the whole space. The edges of the order-$(k + 1)$ diagram are extensions of those of the order k;

7. Let $E^{(k)}(i, j)$ be an edge of the order-k diagram. Hence
$$E^{(k)}(i, j) = O(a, b, .., i) \cap O(a, b, \ldots, j).$$
   The order-$(k - 1)$ diagram gives
$$E^{(k-1)}(i, j) = O(a, b, .., i, j) \cap O(a, b, \ldots, j, i),$$

and this pattern repeats to reach the order-1 diagram

$$E^{(1)}(i,j) = O(i,j) \cap O(j,i);$$

8. The sequence of proximity in a region is determined by using the property 1. The inner half-plane determined by the order-1 edge indicates the closest site point. The order-2 edge indicates the next site point in proximity and this pattern repeats till the order-k;
9. The region $O(X)$ indicates that the closest points belongs to X and the farthest ones belongs to $C - X$;
10. The edges of the order k are extensions of those of the order $k - 1$. Therefore, a vertex is common to the edges of orders k and $k - 1$

$$\bigcap_{i,j \in C} E^k(i,j) \equiv \bigcap_{i,j \in C} E^{k-1}(i,j).$$

6.4 The Ordered Order-k MWVDs

As defined in Section 6.3.1, the ordered order-k Voronoi diagram gives the proximity relations between site points in a sequence of proximity. Then, the multiplicative weighting is combined to yield the ordered order-k MWVD, in which, a Voronoi region, depicted as $O(X)$, $X = \{i, j, \ldots, q\}$, is the locus of all the points closest to the site point c_i.

If c_i is removed, the site point c_j becomes the closest to O and c_q is the farthest one. Each edge $E^{(\xi)}(i,j)$ divides the plane into half-planes $\pi_i^{(\xi)}$, $\pi_j^{(\xi)}$, in which ξ indicates the order of the edge. The Voronoi region is the intersection of the half-planes

$$O(X) = \bigcap_{n \in X, \xi \in [1,k]} \pi_n^{(\xi)}. \tag{6.5}$$

For instance, see the region

$$O(3, 2, 1) = \pi_3^{(1)} \cap \pi_2^{(2)} \cap \pi_1^{(3)}$$

illustrated in Figure 6.4.

6.5 The Path-Loss Prediction

The power loss between a transmitter and a receiver is called path loss. Several models aim at predicting this loss through analytic and measurement based methods. The appropriate, environment-specific, model must be chosen for

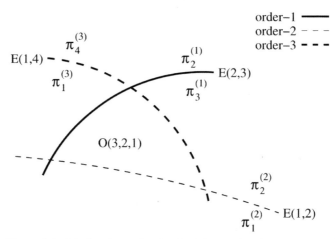

Figure 6.4   Limits of an ordered order-3 Voronoi region in the plane.

accurate path loss prediction. For macro-cells, the most common models are: Lee, Okumura–Hata and COST–Hata (COST-231, 1999). For micro- and pico-cells, the COST–Walfish–Ikegami, the Xia–Bertoni model and ray-tracing method are useful. The predicted path-loss depend on frequency, antenna heights, distance, and propagation environment characteristics. Generically, the path-loss can be expressed as (Hata, 1998)

$$L = a + b \log(d), \tag{6.6}$$

in which a and b are parameters dependent on the model.

6.6 The Mobile Cellular Network as a Voronoi Diagram

The coverage is determined by the received radio signal at the mobile. The received field strength becomes more attenuated far from the BS. Thus, there is a maximum accepted path-loss. When the limit is achieved, the current link is broken and a handoff or blocking operation is executed.

6.6.1 Cell Radius

For an isotropic propagation environment and an omnidirectional antenna, the mean boundary of a cell is a circumference. Its radius depend on the propagation parameters and is obtained from the downlink received power equation

$$P_r = P_t + G_t + G_r - L, \tag{6.7}$$

in which P_t is the BS transmit power, G_t and G_r are the BS and MS gain, respectively, and L is the link BS–MS path-loss. The received power threshold is denoted Z, thus the condition $P_r \geq Z$ defines the cell. Substituting Equation (6.6) into Equation (6.7), yields

$$P_r = P_t + G_t + G_r - a - b\log(d). \tag{6.8}$$

At the cell border $P_r = Z$ and the distance d corresponds to the cell radius

$$r = 10^{\frac{P_t + G_t + G_r - a - Z}{b}}. \tag{6.9}$$

The cell radius can also be estimated by statistical methods in which the signal envelope is considered to follow the Rayleigh, Lognormal, and Suzuki distributions (Yacoub, 1996).

6.6.2 The Two-Cell Model

Consider two adjacent cells as seen in Figure 6.5 and a mobile connected to BS_1. When the mobile is far from BS_1 and approaches BS_2, the system

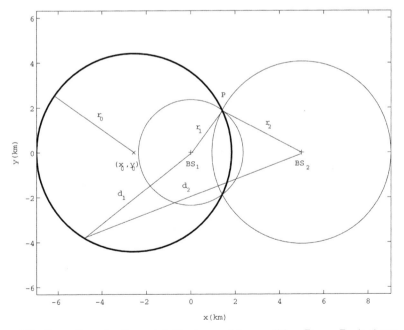

Figure 6.5 Two adjacent cells model. The locus of the condition $P_{r1} = P_{r2}$ is shown as a thick line.

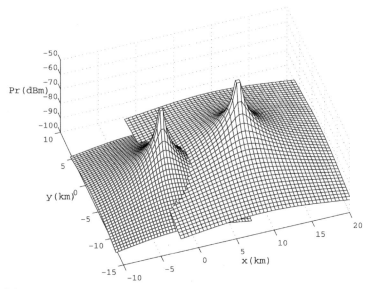

Figure 6.6   Two adjacent cells model in a three-dimension diagram. The locus of the border condition $P_{r1} = P_{r2}$ can be seen as a circular arc between the cells.

analyzes the current, and the alternative, connection quality. If the current connection quality degenerates below a certain threshold, the system changes the connection to BS_2. A proximity rule can be defined based on the received power

IF $P_{r1} \geq P_{r2}$, the mobile is closer to BS_1.
ELSE, the mobile is closer to BS_2,

in which P_{rj} is the downlink received power transmitted by the jth BS. The locus of the condition $P_{r1} = P_{r2}$ is a circumference (shown in Figure 6.5 as a thick line) which represents the border of the two cells. It is illustrated also in three dimensions in Figure 6.6, in which the transmit power decreases according to the Okumura–Hata model.

The center $\langle x_0, y_0 \rangle$ and radius r_0 of the circumference are given as

$$x_0 = \frac{(w_{12})^2 x_2 - x_1}{(w_{12})^2 - 1};$$ (6.10)

$$y_0 = \frac{(w_{12})^2 y_2 - y_1}{(w_{12})^2 - 1};$$ (6.11)

$$r_0 = \left| \frac{d_{12} w_{12}}{(w_{12})^2 - 1} \right|.$$ (6.12)

The circumferences with radii r_1, r_2 intersect at point P. It gives

$$\frac{d_1}{d_2} = \frac{r_1}{r_2}. \tag{6.13}$$

Expressing Equation (6.13) as

$$\frac{d_1}{r_1} = \frac{d_2}{r_2}, \tag{6.14}$$

yields the definition of the proximity rule in Equation (6.4). Let the BSs be represented by site points, thus the weights correspond to the cells radii

$$w_i = r_i \tag{6.15}$$

and the distance ratio is given by the cells radii ratio

$$w_{ij} = \frac{r_i}{r_j}. \tag{6.16}$$

Thus, the MWVD represents the borders of the cells (Portela and Alencar, 2004a), since the link BS–MS is established according to a proximity rule expressed as a function of the received power. The BS works as a site point determining a Voronoi region by means of its radio signal. The border between two adjacent cells is a circular arc (Yacoub, 1993b). For the particular case where $w_i = w_j$ ($r_i = r_j$), the border is a straight line (a circular arc with infinite radius). The distance ratio w_{ij} can be obtained for the whole service area taking cells pairwise. The distance ratio represents the neighboring relationship of the BSs. For two adjacent BSs, the proximity rule is

IF $d_i \leq w_{ij} d_j$ THEN the mobile is closer to BS_i.
ELSE the mobile is closer to BS_j.

The Figure 6.7 shows the edges between two site points in terms of the distance ratio.

6.6.3 Sectored Cells

For a sectored cell, the transmit power is a function of the antenna pattern. If the antenna gain is assumed constant, the cell sector can be represented by a "sector of a circle" whose radius is determined by Equation (6.9). An example is shown in Figure 6.8, where the cell of BS_1 is omnidirectional and those of BS_2 and BS_3 are three-sectored. The neighboring relationship is analyzed

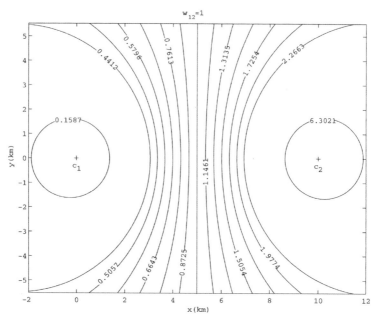

Figure 6.7 Family of edges of two adjacent site points in terms of the distance ratio.

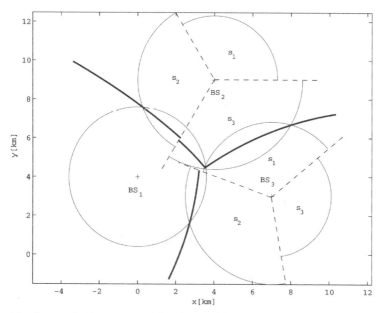

Figure 6.8 Sectored cells represented by a Voronoi diagram. A distance ratio is obtained for each border. The borders are shown as thick lines.

taking adjacent cells pairwise. As an example, consider the border of $BS_2(s_3)$ and $BS_3(s_1)$. The proximity rule defined in Equation (6.4) applied to the border of sectored cells gives the following distance ratio:

$$w_{i(s_p)j(s_q)} = \frac{r_i(s_p)}{r_j(s_q)},$$

(6.17)

in which r is the radius of the sector s (Portela and Alencar, 2004b).

6.6.4 Microcells

A microcell is usually used for hot traffic spots or small coverage holes in a hierarchical structure. A microcell is sometimes underlaid in a macrocell. It can be deployed by using a low transmit power, a low antenna height, or a tilted antenna. This is the case of overlaid coverage of two cells when a cell is inside another. Geometrically, this can be represented by a MWVD (Portela and Alencar, 2005). Figure 6.9 shows an example in which the MWVD principle stated in Equation (6.4) is applied to represent microcells.

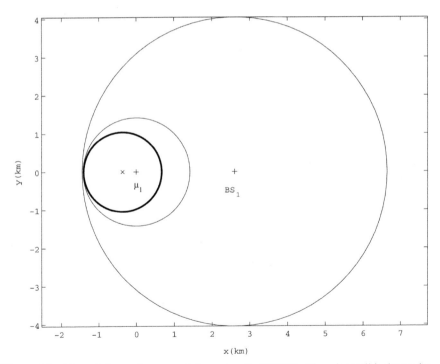

Figure 6.9 Macro/microcell representation using the MWVD. The microcell is denoted μ_1. The border macro/microcell is shown as a thick line.

6.7 Coverage – An Example

The coverage of a cellular network is shown as an example. The Okumura–Hata model has been used for path-loss prediction for a plain terrain, medium sized city with medium trees density, BS heights in the range of 30–200 m, above rooftop, mobile antenna height between 1 and 10 m, a 150–1000 MHz frequency range and a link maximum length of 20 km. The following data is assumed: Antenna gain sum 10 dB ($G_t + G_r = 10$), the mobile antenna height is 3 m and the carrier frequency is 850 MHz. The received power threshold is –90 dBm. The BS location is a technical and financial decision that aims at optimizing the site installation cost. The number of BS is a traffic engineering matter. This example considers realistic data.

6.7.1 Distance Ratio

The distance ratio is obtained using Equation (6.16). The cell radius is obtained from Equation (6.9). The parameters a and b in Equation (6.6) are obtained from the Okumura–Hata path-loss prediction formula:

$$L = 69.55 + 26.16 \log(f) - 13.82 \log(h_b) - a(h_m)$$
$$+ (44.9 - 6.55 \log(h_b)) \log(d), \tag{6.18}$$
$$a(h_m) = (1.1 \log(f) - 0.7)h_m - (1.56 \log(f) - 0.8),$$
$$a = 69.55 + 26.16 \log(f) - 13.82 \log(h_b) - a(h_m),$$
$$b = 44.9 - 6.55 \log(h_b),$$

the antenna height is denoted h_b for BS and h_m for MS, f is the carrier frequency. The BS data are shown in Table 6.1 and the distance ratio in Table 6.2. The cells are represented by the MWVD shown in Figure 6.10. The

Table 6.1 Base Station (BS) data: location, power and antenna heights; Okumura–Hata parameters

BS	Location (km)	Power (dBm)	Antenna Height (m)	Cell Radius (km)	Okumura–Hata Parameters
1	(2,10)	37	55	3.6058	$a_1 = 118.34, b_1 = 33.50$
2	(5,15)	32	65	2.7792	$a_2 = 117.33, b_2 = 33.02$
3	(7,3)	40	61	4.6876	$a_3 = 117.72, b_3 = 33.20$
4	(7,9)	40	56	4.4743	$a_4 = 118.23, b_4 = 33.44$
$5s_1$	(11,14)	37	38.2	3.0000	$a_5s_1 = 120.52, b_5s_1 = 34.53$
$5s_2$	(11,14)	37	60	3.7740	$a_5s_2 = 117.81, b_5s_2 = 33.25$
$5s_3$	(11,14)	40	45.3	4.0000	$a_5s_3 = 119.49, b_5s_3 = 34.05$
6	(12,8)	35	55	3.1423	$a_6 = 118.34, b_6 = 33.50$
μ_1	(4.5,1)	28	46.6	1.8000	$a_{\mu 1} = 119.31, b_{\mu 1} = 33.97$

Table 6.2 The distance ratio between two adjacent cells

w_{12}	w_{13}	w_{14}	w_{24}	$w_{25(s_2)}$	w_{34}
1.2973	0.7691	0.8058	0.6212	0.7364	1.0477
w_{36}	$w_{45(s_2)}$	$w_{45(s_3)}$	w_{46}	$w_{5(s_3)6}$	$w_{3(\mu 1)}$
1.4918	1.1856	1.1185	1.4239	1.2010	2.6042

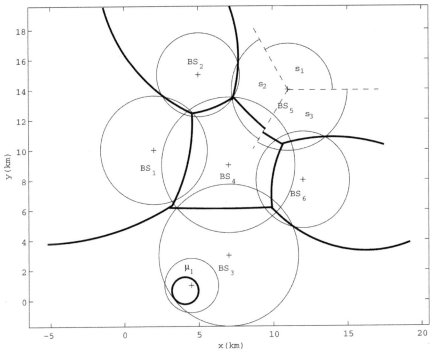

Figure 6.10 A cluster of six BS. The Voronoi diagram is composed of circular arcs shown as thick lines. The cell of BS_5 is three-sectored. The cell of BS_3 has a microcell denoted as μ_1.

circumferences radii represent the weights of the BSs and do not compose the diagram. The Voronoi region contours are circular arcs whose radii are related to the radii of the neighbor cells according to Equations (6.12) and (6.16).

6.8 Spatial Analysis of the Coverage

Some traffic operations including handoff, cell breathing, and blocking are closely related to proximity between BSs. These proximity relations can also have influence in co-channel interference, outage, frequency reuse, and channel allocation scheme. Spatial information can be derived from the cellular Voronoi diagram and used in spatial traffic models.

A cluster of four cells is given as an example. The COST–Hata model is used for path-loss prediction, carrier frequency of 1800 MHz, omnidirectional antennas whose gains sum $G_t + G_r = 9\,dB$, received power threshold of -100 dBm in an urban environment. Further transmission data are given in Table 6.3. The corresponding order-3 diagram is shown in Figure 6.11.

Table 6.3 BS data

BS	Location	Power (dBm)	Antenna Height (m)	Cell Radius (km)	COST–Hata Parameters
1	(1,1)	40	41.56	3.6	$a_1 = 129.81, b_1 = 34.29$
2	(3,10)	40	50.89	4.0	$a_2 = 128.70, b_2 = 33.72$
3	(5,6)	34	73.55	3.2	$a_3 = 126.49, b_3 = 32.67$
4	(9,12)	43	48.86	4.8	$a_4 = 128.94, b_4 = 33.83$

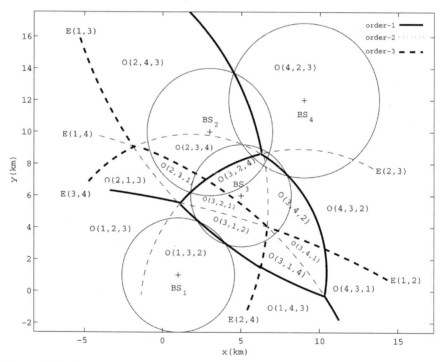

Figure 6.11 Ordered order-3 MWVD, representing four adjacent BSs. Orders 1, 2, and 3 diagrams are superimposed.

Table 6.4 Voronoi edges and distance ratio

	$E(1,2)$	$E(1,3)$	$E(1,4)$	$E(2,3)$	$E(2,4)$	$E(3,4)$
Center	$(-7.5,-37.6)$	$(20,24.8)$	$(-9.3,-13.1)$	$(8.6,-1.1)$	$(-10.6,5.5)$	$(1.8,1.2)$
Radius (km)	43.869	27.131	23.318	9.962	17.205	8.655
Distance ratio	0.9	1.12	0.75	1.24	0.833	0.666

The Voronoi edges are described by center, radius and distance ratio shown in Table 6.4. The spatial analysis is presented as follows.

- The Voronoi regions identify locations proximity: $O(i, j, p)$ means that a handoff may occur primarily between BS_i and BS_j and, secondly between BS_i and BS_p;
- Information about spatial traffic can be explored. For example, if a handoff initiates in a highway from $O(3, 4, 2)$ toward $O(2, 4, 3)$, the expected target station is BS_2 followed by BS_4. In addition, it is expected a high handoff rate between BS_4 and BS_4 because of the high velocity of the mobiles in the highway;
- Cell breathing – Assume BS_i is heavy loaded. When it breathes, its overlapping area changes dynamically and the proximity relations in the neighborhood is altered. For each step in power decreasing of BS_i, all the neighbor edges move in a non-linear manner. These changing affect the load of the neighbor cells and the handoff operations. The transmit power reduction alters the edges according to Equations (6.9), (6.10), (6.11), and (6.12); and the new edge $E^*(i, j)$ is defined in terms of the distance ratio

$$w_{ij} = 10^{\frac{P_{ti}^* + G_{ti} + G_r - a_i - Z}{b_i} - \frac{P_{tj} + G_{tj} + G_r - a_j - Z}{b_j}}, \quad (6.19)$$

  in which $P_{ti}^* = P_{ti} - \Delta P$ is the power of BS_i after a step of power variation ΔP (Jalali, 1998). An example is shown in Figure 6.12 in which BS_2 is breathing. The edges $E(2, j)$ move to $E^*(2, j)$ for a reduction of 5% in r_2;
- All the regions denoted $O(i, \dots)$ identify the coverage of BS_i;
- In a handoff prioritized scheme, if a BS is heavy loaded in a certain time interval, the second closest BS can support the originating calls from the heavy loaded BS;
- The traffic can be planned based on spatial information. For example, BS_3 and BS_4 can support part of the BS_2 traffic in a certain time interval for originating calls from $O(2, 3, 4)$ and $O(2, 4, 3)$. Further, BS_1 can support the originating calls from $O(2, 1, 3)$ and $O(2, 3, 1)$ when BS_2 is heavy loaded;

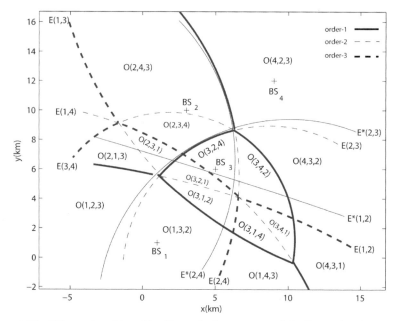

Figure 6.12 BS$_2$ are breathing. The Voronoi edges move and the Voronoi regions change their dimensions in a non-linear fashion.

- Shopping and financial centers, highways, airports, etc., acquired from the Geographic Information System (GIS), can be seen as nodes of demand[1]. The proximity between a node and a BS is valuable for spatial traffic modeling and planning;
- The farthest BS can be identified for planning frequency reuse and channel allocation schemes. For example, an order-4 diagram shows regions of the type $O(i, j, p, q)$, from which BS$_i$ is the nearest station and BS$_q$ is the farthest one; BS$_i$ can support primarily the traffic in O and BS$_q$ can borrow or reuse channels of BS$_i$.

6.9 Outage Contours

The outage is a condition in which a mobile user is completely deprived of service by the system, a service condition below a threshold of acceptable performance (Jones and Skellern, 1995). This situation is caused by

---

[1]Zones of the cell in which the traffic is considered to be uniform and constant in a certain time interval.

co-channel interference plus noise. This is a probabilistic phenomenon, because the interference occurs randomly, when the channel allocation system fails and allocates the same channel to adjacent cells.

The outage occurs when the signal-to-interference plus noise ratio (SINR) falls below a predetermined protection ratio. Inside the outage contour, all the connections are protected from a co-channel interference. Outside, the connections are subject to outage.

6.9.1 Interference Model

A simplified interference model is presented in Figure 6.13. Two adjacent BSs BS_1 and BS_2 are shown, assuming the BS_2 as the interfering source. The interference may occur in the two following ways:

1. From a BS onto a mobile (downlink),
2. From a mobile onto a BS (uplink).

This model considers only the situation in item 1, the MS as an interfering source is not considered. The locus of the ratio d_1/d_2 is the outage contour. Inside it, the downlink is protected from interference. Outside this,

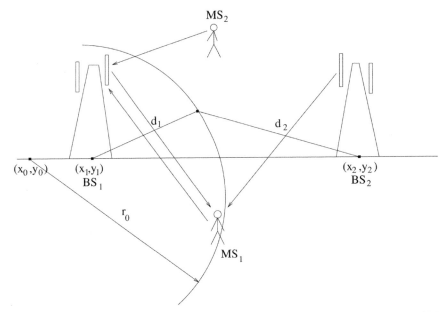

Figure 6.13 Interference model in an arbitrary cellular network. The downlink signal of BS_2 is interfering on a mobile MS_1. The uplink signal of MS_2 is interfering on BS_1.

the downlink is subject to outage occurrence. According to the Apollonius theorem (haruki and Rassias, 1996), the locus of the distance ratio d_1/d_2 is a circumference whose center and radius are given by Equations (6.10), (6.11), and (6.12).

The outage condition can be analyzed by the SINR formula

$$\text{SINR}_1 = 10 \log \left(\frac{p_{r1}}{p_{r2} + n_0} \right) \tag{6.20}$$

in which SINR_1 refers to the interfered downlink, p_{r1} is the received power of the target signal, p_{r2} is the received power of the interfering signal and n_0 is the additive white Gaussian noise (AWGN) in watts. According to the expression (6.20)

IF SINR$\geq \lambda_{\text{th}}$ the mobile is free of outage.
ELSE the mobile is subject to outage,

in which λ_{th} is the protection ratio. It is reasonable to consider $p_{r2} \gg n_0$. Therefore, the SINR_1, in dB, can be treated as SIR_1 and given by

$$\text{SIR}_1 = P_{t1} - L_1 - P_{t2} + L_2, \tag{6.21}$$

in which P_{t1} is the BS_1 transmit power and P_{t2} is the interfering transmit power in dBm. For a given protection ratio λ_{th}, the distance ratio

$$w_{12} = \frac{d_1}{d_2} \tag{6.22}$$

can be used to define the outage contour (circumference of the Apollonius circle). This contour is a bisector dividing the plane into half-planes, each one representing the coverage of the BS. It is also an edge of an MWVD. All the points surrounding BS_1, in which the distance ratio is verified, determine the locus in which the SIR equals λ_{th} and define the outage contour. The MWVD is generated using as input the BS locations $\langle x_i, y_i \rangle$ and the distance ratio w_{12}.

From the Equation (6.21), the expression

$$\lambda_{\text{th}} = P_{t1} + G_{t1} + G_{r1} - a_1 - b_1 \log(d_1)$$
$$- P_{t2} - G_{t2} - G_{r2} + a_2 + b_2 \log(d_2) \tag{6.23}$$

is derived. From Equation (6.23), the distance ratio d_1/d_2 can be numerically computed.

6.9.2 The Distance Ratio

To define the outage contour is necessary to obtain the distance ratio corresponding to the equality in Equation (6.23). The procedure is described as follows.

INPUT: BS antenna height (h_b) and location $\langle x, y \rangle$ transmit power (h_b), protection ratio (λ_{th}).

OUTPUT: d_1, d_2, w_{12}.

Step 1: Initialize d_1, d_2.

Step 2: Compute the parameters of the path loss prediction model: a, b.

Step 3: Compute P_{r1}, P_{r2}.

Step 4: IF $P_{r1} - P_{r2} > \lambda_{th}, d_1 + +, d_2 - -$, GO TO 3.
ELSE $w_{12} = d_1 / d_2$.

Step 5: EXIT.

6.9.3 An Example

Consider a pair of BS with the data shown in Table 6.5. The carrier frequency is 1800 MHz, COST–Hata path loss prediction model, received power threshold of -100 dBm, mobile antenna height of 3 m.

The outage contour is the circumference described by center: $\langle -0.5, 0 \rangle$ and radius: $r_0 = 1.671$ km shown in Figure 6.14. The method to obtain the distance ratio is graphically shown in Figure 6.15.

In a mobile cellular network, the connections are established on a proximity rule basis, which takes into account the connection quality. This rule can be expressed in terms of a weighted distance whose weight depends on the path loss, BS power, antenna heights, and cell radius. The weighted distances generate an MWVD. This representation is applicable to omni, sectored and microcell.

The BS is represented as a site point, in the planning process, whose weight corresponds to the cell radius. The resulting spatial tessellation of the service area is useful to plan coverage, handoff, outage, frequency reuse, and channel allocation.

Table 6.5 Base station data to obtain the outage contour

BS	Location	Power (dBm)	Antenna Height (m)	Cell Radius (km)	COST–Hata Parameters
1	(0,0)	43	45	2.356	$a_1 = 130.34, b_1 = 34.01$
2	(5,0)	45	50	2.806	$a_2 = 129.77, b_2 = 33.77$

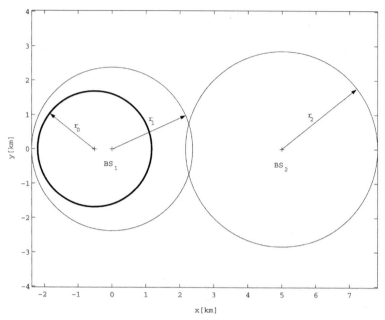

Figure 6.14 The outage contour is a circumference. It is also an edge of a MWVD.

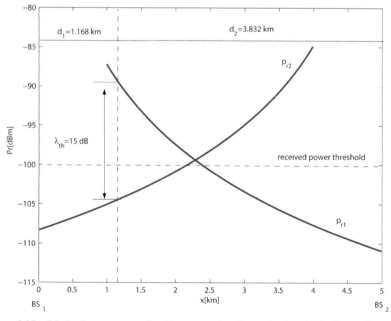

Figure 6.15 Method to compute the distance ratio. The point in which $P_{r1} - P_{r2} = \lambda_{th}$ defines d_1, d_2 and $w_{12} = d_1/d_2$.

Bibliography

[1] Aguiar, J. R., Alencar, M. S. Jr., V. C. R., and Lopes, W. T. A. (2016). "Estudo de Caso na Faixa de Telefonia Móvel Celular (800 MHz), in *Proceedings of the Simpósio Brasileiro de Microondas e Optoeletrônica, Congresso Brasileiro de Eletromagnetismo (MOMAG 2006)*, volume Publicado em CD, CD299, Belo Horizonte, Brasil.

[2] Aguiar, J. R., de Alencar, M. S., and da Rocha Jr. V. C. (2004). "Interferência Eletromagnética na Faixa de Telefonia Móvel Celular," in *Proceedings of the Anais do Simpósio Brasileiro de Microondas e Optoeletrônica*, volume Publicado em CD, São Paulo, Brasil.

[3] Alencar, M. S. (2000). *Telefonia Digital, Terceira Edição*. São Paulo: Editora Érica Ltda.

[4] Alencar, M. S. (2011). *Telefonia Digital, Quinta Edição*. São Paulo: Editora Érica Ltda.

[5] Alencar, M. S. (2012). *Telefonia Celular Digital, Terceira Edição*. São Paulo: Editora Érica Ltda.

[6] Alencar, M. S., and da Rocha Jr. V. C. (2005). *Communication Systems*. Boston, MA: Springer.

[7] Allen, S., Hurley, S., Taplin, R., and Whitaker, R. (2001). "Automatic cell planning of broadband fixed wireless networks," in *Proceedings of the 53rd IEEE Vehicular Technology Conference – VTC 2001 Spring, Rhodes, Greece*, volume 4.

[8] Andersen, J. B., Rappaport, T. S., and Yoshida, S. (1995). Propagation measurements and models for wireless communications channels. *Proc. IEEE Commun. Mag.* 33(1), 42–49.

[9] Assis, M. S. (1994). *Telefonia Móvel Celular: Situação Atual e Perspectivas Futuras*. Relatório Técnico, Ministério das Comunicações, Secretaria de Serviços de Comunicações.

[10] Astély, D., Dahlman, E., Furuskär, A., Jading, Y., Lindström, M., and Parkvall, S. (2009). LTE: The evolution of mobile broadband. *J. IEEE Commun. Magazine* 47, 44–51.

[11] Aurenhammer, F. (1987). Power diagrams: properties, algorithms and applications. *SIAM J. Comput.* 16, 78–96.

[12] Aurenhammer, F. (1991). Voronoi diagrams – A survey of a fundamental geometric data structure. *ACM Comput. Surv.* 23, 345–405.

[13] Basch, J., Guibas, L., and Zhang, L. (1997). "Proximity problems on moving points," in *Proceedings of the 13th Annual ACM Symposium on Computational Geometry*, 344–351.

[14] Bello, P. A. (1963). Characterization of randomly time-variant linear channels. *IEEE Trans. Commun. Syst.* 360–393.

[15] Blahut, R. E. (1999). *Digital Transmission of Information*. Reading, MA: Addison-Wesley Publishing Co.

[16] Blake, I. F. (1987). *An Introduction to Applied Probability*. Malabar, FL: Robert E. Krieger Publishing Co.

[17] Bultitude, R. J. C. (1987). Measurement, characterization and modeling of indoor 800/900 MHz radio channels for digital communications. *IEEE Commun. Magazine* 25(6), 5–12.

[18] Campelo, F., Guimaraes, F., and Igarashi, H. (2007). "Overview of artificial immune systems for multi-objective optimization," in *Proceedings of the 4th International Conference on Evolutionary Multi-Criterion Optimization, EMO 2007,* eds S. Obayashi, K. Deb, C. Poloni, T. Hiroyasu, and T. Murata, (Berlin: Springer), 937–951.

[19] Castro, L. and Timmis, J. (2002). *Artificial Immune Systems: A New Computational Intelligence Approach*. Berlin: Springer.

[20] Cheung, J. C. S., Beach, M. A., and McGeehan, J. P. (1994). Network planning for the third-generation mobile radio systems. *IEEE Commun. Mag.* 54–59.

[21] CNTr (1992). *Básico de Telefonia Móvel Celular*. Brasil: Apostila, Centro Nacional de Treinamento da Telebrás.

[22] Coello, C., Lamont, G., and Veldhuizen, D. V. (2007). *Evolutionary Algorithms for Solving Multi-Objective Problems*. Berlin: Springer.

[23] Cortes, N., and Coello, C., (2005). Solving multiobjective optimization problems using an artificial immune system. *Genetic Program. Evolvable Mach.* 6, 163–190.

[24] COST-231 (1999). *Digital Mobile Radio towards Future Generation Systems*. Final report, European cooperation in the field of scientific and technical research.

[25] Dahlman, E., Parkvall, S., Skold, J., and Beming, P. (2008). *3G Evolution: HSPA and LTE for Mobile Broadband,* 2nd Edn. Waretown, NJ: Academic Press.

[26] de Oliveira, J. N. C. (2004). *Estudo de Modelos de Predição para Telefonia Móvel Celular*. Dissertação de Mestrado, Universidade Federal de Pernambuco.

[27] de Oliveira, J. N. C., Alencar, M. S., da Rocha Jr. V. C., and Lopes, W. T. A. (2006). "A new propagation model for cellular planning," in *Proceedings of the International Telecommunications Symposium (ITS 2006)*, volume Publicado em CD, Fortaleza, Brasil.

[28] Devasirvatham, D. M. J. (1984). Time delay measurements of wideband radio signals within a building. *Electron. Lett.* 20(23), 950–951.

[29] Dhir, A. (2004). *The Digital Consumer Technology Handbook*. Burlington, VT: Elsevier.

[30] Falconer, D. D., Adachi, F., and Gudmundson, B. (1995). Time division multiple access methods for wireless personal communications. *IEEE Commun. Magazine*, 33(1), 50–57.

[31] Freeburg, T. A. (1991). Enabling technologies for wireless in-building network communications – four technical challenges, four solutions. *IEEE Commun. Magazine* 29(4), 58–64.

[32] Freschi, F. (2006). VIS: an artificial immune network for multi-objective optimization. *Eng. Optim.* 38(8), 975–996.

[33] Gagliardi, R. M. (1988). *Introduction to Communications Engineering*. New York, NY: Wiley.

[34] Gersho, A. (1969). Adaptive equalization of highly dispersive channels for data transmission. *Bell System Technical J.* 1, 55–71.

[35] Gomes, A. M. N. (2001). *Análise do Comportamento em Campo Próximo para Antenas de Comunicações Móveis*. Dissertação de Mestrado, Universidade Federal da Paraba.

[36] Guimarães, F., Palhares, R., Campelo, F., and Igarashi, H. (2007). Design of mixed control systems using algorithms inspired by the immune system. *Inform. Sci.* 177, 4368–4386.

[37] Hao, Q., Soong, B., Ong, J., Soh, C., and Li, Z. (1997). A low-cost cellular mobile communication system: a hierarquical optimization network resource planning approach. *IEEE J. Sel. Areas Commun.* 15(7), 1315–1326.

[38] Hao, Q., Soong, B.-H., Gunawan, E., Ong, J.-T., Soh, C.-B. and Li, Z. (1997). A low-cost cellular mobile communication system: a hierarchical optimization network resource planning approach. *IEEE J. Sel. Areas Commun.* 15(7), 1315–1326.

[39] Haruki, H., and Rassias, T. M. (1996). A new characteristic of Möbius transformations by use of Apollonius points of triangles. *J. Math. Analysis Appl.* 197(1), 14–22.

[40] Hashemi, H. (1991). "Principles of digital indoor radio propagation,"in *Proceedings of the IASTED International Symposium on Computers, Electronics, Communication and Control*, Calgary, AB, Canada, 271–273.

[41] Hata, M. (1998). "Propagation loss prediction models for land mobile communications," in *Proceedings of the International Conference on Microwave and Millimeter Wave Technology (ICMTT 98)*.

[42] Held, M., and Williamson, R. B. (2004). Creating electrical distribution boundaries using computational geometry. *IEEE Trans. Power Systems* 19(3), 1342–1347.

[43] Holma, H., and Toskala, A. (2009). *LTE for UMTS OFDMA and SC-FDMA Based Radio Access*. Finland: John Wiley & Sons Ltd.

[44] Huang, X., Behr, U., and Wiesbeck, W. (2000). "Automatic cell planning for a low-cost and spectrum efficient wireless network" in *Proceedings of the Global Telecommunications Conference, GLOBECOM '00*, San Francisco, CA, USA.

[45] Jalali, A. (1998). "On cell breathing in CDMA networks," in *Proceedings of the IEEE International Conference on Communications (ICC'98)*, volume 2, 985–988.

[46] Jones, B., and Skellern, D. J. (1995). "Outage contours and cell size distributions in cellular and microcellular networks," in *Proceedings of the IEEE 45th Vehicular Technology Conference*, volume 1, 145–149.

[47] Kennedy, R. S. (1969). *Fading Dispersive Communication Channels*. New York, NY: Wiley-Interscience.

[48] Khan, F. (2009). *LTE and the Evolution to 4G Wireless,* 1st edn. New York, NY: Moray Rummey.

[49] Lafortune, J.-F. and Lecours, M. (1990). Measurement and modeling of propagation losses in a building at 900 MHz. *IEEE Trans. Veh. Technol.* 39, 101–108.

[50] LaFortune, J.-F., and Lecours, M. (1990). Measurement and modeling of propagation losscs in a building at 900 MHz. *IEEE Trans. Veh. Technol.* 39(2).

[51] Lecours, M., Chouinard, J.-Y., Delisle, G. Y., and Roy, J. (1988). Statistical modeling of the received signal envelope in a mobile radio channel. *IEEE Trans. Veh. Technol.* 37, 204–212.

[52] Lee, D.-T. (1982). On k-nearest neighbor Voronoi diagrams in the plane. *IEEE Trans. Comput.* C-31:478–487.

[53] Lee, W. C. Y. (1989). *Mobile Cellular Telecommunications Systems*. New York, NY: McGraw-Hill Book Company.

[54] Lee, W. C. Y. (1990). Estimate of channel capacity in rayleigh fading environment. *IEEE Trans. Veh. Technol.* 39, 187–189.

[55] Lee, W. C. Y. (1991). Overview of cellular CDMA. *IEEE Trans. Veh. Technol.* 40, 291–302.

[56] Lee, W. C. Y. (1993). *Mobile Communications Design Fundamentals.* New York, NY: John Wiley & Sons Ltd.

[57] Lee, W. C. Y. (1992). *Mobile Cellular Telecomunications.* New York, NY: McGraw-Hill Book Company.

[58] Leon-Garcia, A. (1989). *Probability and Random Processes for Electrical Engineering.* Reading, MA: Addison-Wesley Publishing Co.

[59] Luh, G. and Chueh, C. (2004). Multi-objective optimal design of truss structure with immune algorithm. *Comput. Struct.* 82:829–844.

[60] Macario, R. (1991). *Modern Personal Radio Systems.* London: The Institution of Electrical Engineers.

[61] Macchi, C., Jouannaud, J.-P., and Macchi, O. (1975). Récepteurs adaptatifs pour transmission de données a grande vitesse. *Ann. Télécomm.* 30, 311–330.

[62] Mishra, A. R. (2007). *Advanced Cellular Network Planning and Optimization.* West Sussex, England: John Wiley & Sons, Ltd.

[63] Nanda, S. and Goodman, D. J. (1992). *Third Generation Wireless Information Networks.* Boston, MA: Kluver Academic Publishers.

[64] Newman, D. B. Jr. (1986) FCC authorizes spread spectrum. *IEEE Commun. Mag.* 24, 46–47.

[65] Okumura, Y., Ohmori, E., Kawano, T., and Fukuda, K. (1968). Field strenght and its variability in VHF and UHF land-mobile radio service. *Rev. Electr. Commun. Lab* 16, 825–873.

[66] Pahlavan, K., Ganesh, R., and Hotaling, T. (1989). Multipath propagation measurements on manufacturing floors at 910 MHz. *Electron. Lett.* 25, 225–227.

[67] Portela, J. N. and Alencar, M. S. (2004a). "Outage contour using a Voronoi diagram," in *Proceedings of the IEEE Wireless Communication and Networking Conference (WCNC'04)*, Atlanta, GA. 21–26.

[68] Portela, J. N. and Alencar, M. S. (2004b). "The mobile cellular network as a set of voronoi diagrams," in *Proceedings of the XXI Simpósio Brasileiro de Telecomunicações, (SBrT'04)*, Belem, 6–9.

[69] Portela, J. N. and Alencar, M. S. (2005). "Spatial analysis of the overlapping cell area using voronoi diagrams," in *Proceedings of the International Microwave and Optoelectronics Conference (IMOC 2005)*, Brasilia, 25–28

[70] Proakis, J. G. (1990). *Digital Communications.* New York, NY: McGraw-Hill Book Company.

[71] Qualcomm, (1992). *The CDMA Network Engineering Handbook*, vol. 1. Qualcomm Incorporated, San Diego, CA.

[72] Qureshi, S. U. H. (1985). Adaptive equalization. *Proc. IEEE* 73, 1349–1387.

[73] Raisanen, L. (2006). *Multi-Objective Site Selection and Analysis for GSM Cellular Network Planning*. Ph.D. thesis, Cardiff University, Cardiff.

[74] Raisanen, L. (2007). A permutation-coded evolutionary strategy for multi-objective GSM network planning. *J. Heuristics*.

[75] Rappaport, T. S. (1989). Indoor radio communications for factories of the future. *IEEE Commun. Mag.* 2715–24.

[76] Rawnsley, R. and Hurley, S. (2000). "Towards automatic cell planning." in *The 11th IEEE International Symposium on Personal, Indoor and Mobile Radio Communications, 2000 – PIMRC 2000,* London, 1583–1588.

[77] Ring, D. H. (1947). Mobile telephony – wide area coverage – Case 20564. Memorandum: Bell Telephone Laboratories, Inc.

[78] Robertazzi, T. G., (1998). *Planning Telecommunication Networks*. Rome: IEEE Press.

[79] Roessler, A., Schilienz, J., Merkel, S., and Kottkamp, M. (2014). Lte-advanced (3gpp rel.12) technology introduction.

[80] Rouault, J. M. (1976). *Télétrafic"*. Paris: Éditions Eyrolles.

[81] Saleh, A. A. M. and Valenzuela, R. A. (1987). A statistical model for indoor multipath propagation. *IEEE J. Sel. Areas Commun.* 5, 128–137.

[82] Schwartz, M. (1970). *Information Transmission, Modulation, and Noise*. New York, NY: McGraw-Hill.

[83] Schwartz, M., Bennett, W., and Stein, S. (1966). *Communication Systems and Techniques*. New York: McGraw-Hill.

[84] Shamos, M. and Hoey, D. (1975). "Closest-point problems," in *Proceedings of the Annual IEEE Foundations of Computer Science*, Rio de Janeiro: Edgard Blucher Ltda., 151–162.

[85] Shepherd, N. H. (ed.) (1988). Received signal fading distribution. *IEEE Trans. Veh. Technol.* 37, 57–60.

[86] Steele, R. (1993). Speech codecs for personal communications. *IEEE Commun. Mag.* 31, 76–83.

[87] Stuber, G. L. (1991). *Principles of Mobile Communication*. Boston, MA: Kluwer Academic Publishers.

[88] Thom, D. (1991). *Characterization of indoor wireless channel in the presence of multipath fading*. Report 1, Waterloo, ON: University of Waterloo.

[89] Valenzuela, R. A. (1989). Performance of adaptive equalization for indoor radio communications. *IEEE Trans. Commun.* 37, 291–293.

[90] Vasquez, M. and Hao, J.-K. (2001). A heuristic approach for antenna positioning in cellular networks. *J. Heuristics* 7:443–472.

[91] Xia, H. H. (1996). An analytical model for prediction path loss in urban and suburban environments. *PIMRC96*, 2(3):99.

[92] Yacoub, M. D. (1993a). *Foundations of Mobile Radio Engineering*. Boca Raton, FL: CRC Press.

[93] Yacoub, M. D. (1993b). *Foundations of Mobile Radio Engineering*. New York, NY: CRG Press.

[94] Yacoub, M. D. (1996). *The Mobile Communications Handbook*. New York, NY: CRG Press.

[95] Yegani, P. and Mcgillen, C. D. (1991). A statistical model for the factory radio channel. *IEEE Trans. Commun.* 29, 1445–1454.

Index

About the Authors

Marcelo Sampaio de Alencar was born in Serrita, Brazil in 1957. He received his Bachelor Degree in Electrical Engineering, from Universidade Federal de Pernambuco (UFPE), Brazil, 1980, his Master Degree in Electrical Engineering, from Universidade Federal da Paraiba (UFPB), Brazil, 1988, and his Ph.D. from University of Waterloo, Department of Electrical and Computer Engineering, Canada, 1994. Marcelo S. de Alencar has more than 30 years of engineering experience, and 24 years as an IEEE Member, currently as Senior Member. For 18 years he worked for the Department of Electrical Engineering, Federal University of Paraiba, where he was Full Professor and supervised 46 graduate and several undergraduate students.

During that period he was responsible for the creation and administration of three communication and signal processing laboratories in his University, management of several contracts, for consulting and project development, with many companies, agencies, and universities, including Telecom Italia Mobile, MCI-Embratel, Telebras, Chesf, CNPq (Brazilian Council for Scientific Research and Development), Finep (Federal Government Agency to Fund Studies and Projects), UFPE, UFMA, Siemens, Contol, Oi, CSU, Unicamp and Telern. He is member of the Consultative Committee on Digital Television of the Ministry of Communications and member of the Technical Chamber for the Development of Telecommunications (Finep). He was responsible for the development of the first courses on Mobile Cellular Communications and Digital Television in the North and Northeast regions of Brazil.

Since 2003, he is Chair Professor at the Department of Electrical Engineering, Federal University of Campina Grande, Brazil. Previously, he worked, between 1982 and 1984, for the State University of Santa Catarina (UDESC). He spent some time working for MCI-Embratel and University of Toronto, as Visiting Professor. He is founder and President of the Institute for Advanced Studies in Communications (Iecom), founder and Vice-President of the Academy of Sciences of Paraiba (ACParaiba). He has been awarded several scholarships and grants, including three scholarships and several research grants from the Brazilian National Council for Scientific and Technological

Research (CNPq), a scholarship from the University of Waterloo, a scholarship from the Federal University of Paraiba, an achievement award for contributions to the Brazilian Telecommunications Society (SBrT), an award from the Medicine College of the Federal University of Campina Grande (UFCG) and an achievement award from the College of Engineering of the Federal University of Pernambuco, during its 110th year celebration. His biography is included in the following publications: Whos Who in the World and Whos Who in Science and Engineering, by Marquis Whos Who, New Providence, USA. Marcelo S. Alencar is a laureate of the 2014 Attilio Giarola Medal.

He published over 350 engineering and scientific papers and 19 books, which include: Digital Television Systems, by Cambridge, Communication Systems, by Springer, Probability Theory, by Momentum Press, Informacao, Codificacao e Seguranca de Redes, by Elsevier, Information Theory, by Momentum Press, Engenharia de Redes de Computadores, Teoria de Conjuntos, Medida e Probabilidade, Ondas Eletromagneticas e Teoria de Antenas, Probabilidade e Processos Estocasticos, Telefonia Celular Digital, Telefonia Digital, Televisao Digital and Sistemas de Comunicacoes, by Editora Erica Ltda, Historia, Tecnologia e Legislacao de Tecomunicacoes, Sexo Conexo, Divulgacao Cientifica, Historia da Comunicacao no Brasil, Solucos da Alma, by Epgraf, Principios de Comunicacoes, by Editora Universitaria da UFPB. He also wrote eight chapters, including one to the book Communications, Information and Network Security, by Kluwer Academic Publishers, and the essay Historical Evolution of Telecommunications in Brazil for the IEEE Foundation.

Marcelo S. de Alencar has contributed in different capacities to the following scientific journals: Editor of the Journal of the Brazilian Telecommunication Society; Member of the International Editorial Board of the Journal of Communications Software and Systems (JCOMSS), published by the Croatian Communication and Information Society (CCIS); Member of the Editorial Board of the Journal of Networks (JNW), published by Academy Publisher; Editor-in-Chief of the Journal of Communication and Information Systems, special joint edition of the IEEE Communications Society (ComSoc) and SBrT. He was member of the SBrT-Brasport Editorial Board. He served as member of the IEEE Communications Society Sister Society Board and as Liaison to Latin America Societies. He also served on the Board of Directors of IEEEs Sister Society SBrT. He is a Registered Professional Engineer. He is a columnist for the traditional Brazilian newspaper Jornal do Commercio, since April 2000, and is currently Vice-President External Relations of SBrT.

His webpage: www.difusaocientifica.com. Head of the Smart Cities Project at the Iecom.

He has been involved as a volunteer with several IEEE, SBrT, SBMO, and other organizations' activities, including being a member of the Advisory or Technical Program Committee in editions of the following events: LATINCOM 2016, SBrT 2016, SBPC 2016, IMOC 2015, APPEIC'2015, CBQEE 2015, ICCMIT 2015, SBrT 2015, SoftCommm 2014, EMC 2014, EUSIPCO 2014, ICACCI 2014, MOMAG 2014, ICED 2014, ICT 2014, IV SIREE, ICWiSe2013, SBrT 2013, RFM 2013, IWT 2013, ICT-EurAsia 2013, SoftComm 2012, MOMAG 2012, SBrT 2012, ICACCI-2012, ISWCS 2012, CBA 2012, PIMRC'11, ISWCS 2011, SoftComm 2011, I2TS 2011, IWT 2010, I2TS 2010, CBEB'2010, MOMAG 2010, ITS 2010, CBA 2010, IEEE WCNC 2010, WPMC'2010, IEEE ISPLC 2009, ITU-T Kaleidoscope 2009, VTC'2009,

WCNC'2009, ICT'09, WCNC 2008, AICT'2008, IMOC 2007, IST 2007, UBICOMM 2007, GIIS 2007, 20th ITC, AICT'2007, SBrT'07, ICC'2007, IWT07, IWCC 2006, IWCMC'2006, ICT'2006, XXII SBT, XXXIII COBENGE, AICT'06, IEEE PIMRC'06, ITS'06, ISSSTA'06, ICT'2005, ICT'2004, PIMRC'05, AICT 2005, SBT 2005, IEEE WSPAWC 2004, IWT'04, IEEE ICC'04, IEEE PIMRC'04, XXI SBT, IEEE WCNC'04, IEEE GLOBECOM 2003, IMOC 2003, XX SBT, CBA 2002, ITS'2002, IEEE PIMRC'02, SBMO 2002, WPMC'02, IEEE ICC'2002, IEEE ICC'2001, IEEE GLOBECOM'01, IEEE GLOBECOM'00, IEEE VTC'00, CBA 2000, IEEE GLOBECOM'99, XVIII SBT, IEEE ICC'99, SBRC'99, IEEE GLOBECOM'98, SBrT/IEEE ITS'98, XVII SBT, IEEE GLOBECOM'93, XIII SBT, TELEMO'96, XVI SBT, XV SBT.

He served as member of the IEEE Communications Society Sister Society Board and as Liaison to Latin America Societies. He also served on the Board of Directors of IEEE's Sister Society SBrT. He is a Registered Professional Engineer and recipient of a grant from the IEEE Foundation. He has been acting as reviewer for several scientific journals, including the following: IEEE Transactions on Communication, IEEE Transactions on Vehicular Technology, IEEE Transactions on Biomedical Engineering, IEEE Communications Letters, Wireless Personal Communications, published by Kluwer Academic Publishers, Journal of the Brazilian Telecommunications Society.

Marcelo S. de Alencar served on the Board of Directors of the Brazilian Telecommunications Society (SBrT), as Vice-President External Relations, for two decades. He also served on the Council of the Brazilian Microwave and Optoelectronics Society (SBMO). He is member of the Institute of

Electronics, Information and Communication Engineering (Japan), member of the Brazilian Microwave and Optoelectronics Society, member of SBPC (Brazilian Society for the Advancement of Science) and member of SBEB (Brazilian Society for Biomedical Engineering).

Djalma de Melo Carvalho Filho was born in Palmeira dos Indios, Brazil in 1972. He received his Bachelor's Degree in Electrical Engineering, from the Federal University of Paraíba (UFPB), Brazil in 1996, his Master's Degree in Electrical Engineering, from Universidade Federal de Campina Grande (UFCG), Brazil in 2000, his Ph.D. in Electrical Engineering from Universidade Federal de Campina Grande (UFCG), Brazil in 2008 and his Bachelor's Degree in Law, from Faculdade de Ciencias Sociais Aplicadas (FACISA), Brazil in 2012.

Djalma de Melo Carvalho Filho has over 15 years of engineering experience. While studying at the Federal University of Campina Grande he took part in many projects involving major engineering companies, such as Siemens and Chesf. He is a founding member of the Institute for Advanced Studies in Communications (IECOM).

Djalma de Melo Carvalho Filho is currently a Professor at the State University of Paraíba (UEPB) and at Faculdade de Ciências Sociais Aplicadas (FACISA). He has taught courses in Computer Networks, Safety Analysis, Computer Networks, Digital Law and Electrical Safety of Electromedical devices. He is also a researcher at the Center for Strategic Technologies in Health (NUTES) and is a Professor at the Postgraduate Programme in Science and Technology in Health (NUTES/UEPB).